Industrial Alumina Chemicals

Industrial Alumina Chemicals

Chanakya Misra
Alcoa Technical Center

ACS Monograph 184

American Chemical Society
Washington, DC   1986

Library of Congress Cataloging-in-Publication Data

Misra, Chanakya, 1937–
  Industrial alumina chemicals.
  (ACS monograph, ISSN 0065-7719; 184)

  Bibliography: p.
  Includes index.

  1. Aluminum oxide. 2. Aluminum hydroxide.
  I. Title. II. Series.
TP245.A4M57   1986   661'.067   85-28587
ISBN 0-8412-0909-X

Copyright © 1986

American Chemical Society

All Rights Reserved. The appearance of the code at the bottom of the first page of each chapter in this volume indicates the copyright owner's consent that reprographic copies of the chapter may be made for personal or internal use or for the personal or internal use of specific clients. This consent is given on the condition, however, that the copier pay the stated per copy fee through the Copyright Clearance Center, Inc., 21 Congress Street, Salem, MA 01970, for copying beyond that permitted by Sections 107 or 108 of the U.S. Copyright Law. This consent does not extend to copying or transmission by any means—graphic or electronic—for any other purpose, such as for general distribution, for advertising or promotional purposes, for creating a new collective work, for resale, or for information storage and retrieval systems. The copying fee for each chapter is indicated in the code at the bottom of the first page of the chapter.

The citation of trade names and/or names of manufacturers in this publication is not to be construed as an endorsement or as approval by ACS of the commercial products or services referenced herein; nor should the mere reference herein to any drawing, specification, chemical process, or other data be regarded as a license or as a conveyance of any right or permission, to the holder, reader, or any other person or corporation, to manufacture, reproduce, use, or sell any patented invention or copyrighted work that may in any way be related thereto. Registered names, trademarks, etc., used in this publication, even without specific indication thereof, are not to be considered unprotected by law.

PRINTED IN THE UNITED STATES OF AMERICA

FOREWORD

ACS MONOGRAPH SERIES was started by arrangement with the interallied Conference of Pure and Applied Chemistry, which met in London and Brussels in July 1919, when the American Chemical Society undertook the production and publication of Scientific and Technological Monographs on chemical subjects. At the same time it was agreed that the National Research Council, in cooperation with the American Chemical Society and the American Physical Society, should undertake the production and publication of Critical Tables of Chemical and Physical Constants. The American Chemical Society and the National Research Council mutually agreed to care for these two fields of chemical progress.

The Council of the American Chemical Society, acting through its Committee on National Policy, appointed editors and associates to select authors of competent authority in their respective fields and to consider critically the manuscripts submitted. The first Monograph appeared in 1921. Since 1944 the Scientific and Technological Monographs have been combined in the Series.

These Monographs are intended to serve two principal purposes: first, to make available to chemists a thorough treatment of a selected area in a form usable by persons working in more or less unrelated fields so that they may correlate their own work with a larger area of physical science; and second, to stimulate further research in the specific field treated. To implement this purpose, the authors of Monographs give extended references to the literature.

ABOUT THE AUTHOR

Chanakya Misra obtained his M.S. in applied chemistry from Calcutta University and in chemical engineering from the Indian Institute of Technology, Kharagpur, India. He completed his doctorate work at the University of Queensland, Australia, in 1970 after working for several years in the alumina industry. Following his doctorate work, he worked in the alumina industry in West Germany and Suriname (South America).

Dr. Misra is presently a senior technical specialist at Alcoa Technical Center, Pittsburgh, PA. His current work involves development of new alumina production processes and products. He holds three U.S. and foreign patents and is the author of eight publications relating to the alumina industry.

CONTENTS

Preface, **ix**

1. Introduction, **1**

2. Aluminum Oxide and Hydroxides, **7**
   Historical, **7**
   Aluminum Hydroxides, **8**
   Aluminum Oxide-Hydroxide (AlOOH), **17**
   Gelatinous Aluminum Hydroxides, **19**
   Aluminum Oxide (Corundum), **22**
   β-Alumina and Related Compounds, **23**
   Phase Relationships in the Al_2O_3–H_2O System, **25**
   Nomenclature, **27**

3. Industrial Production of Aluminum Hydroxides, **31**
   Industrial Production, **31**
   Bauxite: The Principal Raw Material, **32**
   Bayer Process, **33**
   Sinter Process, **38**
   Combination Process, **40**
   Bayer Hydroxide Product Variants, **41**
   Production of Other Aluminum Hydroxides, **46**

4. Aluminum Hydroxides: Industrial Applications, **55**
   Filler in Plastics and Polymer Systems, **55**
   Paper Applications, **63**
   Raw Material for Production of Other Aluminum Compounds, **65**
   Cosmetics and Pharmaceuticals, **67**
   Glass Industry, **67**
   Other Applications, **67**
   Analytical Procedures for Aluminum Hydroxides, **68**

5. Activated Aluminas, **73**
   Thermal Dehydration of Aluminum Hydroxides and Formation
     of Transition Aluminas, **73**
   Thermal Effects, **83**
   Textural Changes: Development of Pore Structure and Surface Area, **86**
   Other Sources of Transition Aluminas, **93**
   Concluding Remarks, **93**

6. Industrial Production of Activated Aluminas, **97**
   Activated Bauxites, **97**
   Activated Aluminas from Bayer Hydroxide, **97**
   Activated Alumina from Boehmite and Gelatinous Aluminum Hydroxide, **102**
   Activated Alumina from Bayerite, **105**

7. Activated Alumina: Adsorbent Applications, **107**
   Gas Drying, **108**
   Liquid Drying, **119**
   Water Purification, **120**
   Hydrocarbon Recovery and Selective Adsorption Applications, **121**
   Maintenance of Power System Oils, **122**
   Chromatography, **123**
   Recent Developments in Adsorbent Aluminas, **126**
   Analytical Methods for Activated Aluminas, **128**

8. Catalytic Aluminas, **133**
   Surface Structure and Catalytic Activity, **134**
   Examples of Commercial Catalytic Processes Using Alumina Catalysts, **138**
   Alumina as Catalyst Support, **143**
   Examples of Catalytic Processes Using Alumina-Supported Catalysts, **144**

9. Sodium Aluminate, **151**
   Physical and Chemical Properties, **151**
   Commercial Production, **153**
   Analysis, **154**
   Uses, **154**

10. Economic Data for Alumina Chemicals, **157**
    Overview, **157**
    Producers, **157**
    Aluminum Hydroxides, **158**
    Adsorbent Activated Aluminas, **159**
    Alumina Catalysts and Catalyst Supports, **159**
    Sodium Aluminate, **160**

Index, **161**

PREFACE

THE INDUSTRIAL USES OF ALUMINA CHEMICALS have grown considerably during the past two decades. Many uses, such as the application of aluminum hydroxide as a fire-retardant filler in plastics, have seen very rapid growth and have been intensively explored both technically and commercially. Activated alumina, though an established product, is currently the subject of renewed interest as an adsorbent. Research aimed at controlling pore structure and surface properties of activated aluminas has contributed to their wide and rapidly developing uses as catalysts and catalyst supports.

The name "alumina" has been commonly used for a broad range of products derived from aluminum hydroxides and products of their thermal decomposition. Their nature, and consequently their uses, is closely related to the structure of the hydroxides and their behavior during thermal decomposition. These topics are thoroughly discussed in this book to emphasize opportunities available for the development of new alumina products and uses.

The main aim of this volume is to provide a comprehensive overview of industrially significant alumina chemicals. This overview covers their origin, properties, manufacture, and uses. Many developments that are technically interesting but that have not yet reached commercial maturity have been briefly discussed. Vigorous research activity in these areas has resulted in a large number of publications. I apologize for not including many significant recent publications in the literature cited; my only excuse is that they have yet to pass the test of commercial implementation.

I thank my employer, Aluminum Company of America, for financially supporting the preparation of the manuscript and making available many of the illustrations. I am particularly thankful to F. S. Williams, Technical Manager, Alumina and Chemicals Division; L. K. Hudson, Technical Director (retired); and Dr. Karl Wefers, Fellow, all of Alcoa Technical Center for their encouragement, support, and technical guidance during the preparation of the manuscript. I also thank Karen Tipinski, Denise Elliott, and Annie Laurie Irwin, secretarial staff at the Alcoa Technical Center, for patiently typing a difficult manuscript.

CHANAKYA MISRA
Alcoa Technical Center
Alcoa Center, PA 15069

August 9, 1985

1
Introduction

The broad term alumina embraces a large number of products having a wide variety of properties and applications.

Chemical and medicinal usage of aluminous materials (alum) goes as far back as the Greek and Roman civilizations. The discovery of the aluminum-rich ore bauxite in 1821 provided a convenient source of aluminum for the preparation of alum, which was the principal aluminum compound of commercial value at that time. Industrial production of pure aluminum hydroxide, used for the manufacture of high-quality aluminum sulfate and alum, followed the development of the sinter process by Le Chatelier in 1875. The demand for pure alumina increased sharply when it became the raw material for aluminum metal production upon development of the Hall–Heroult electrolytic cell in 1886. In the years following (1887–1888), Karl Joseph Bayer developed the bauxite refining process that bears his name. The Bayer process, because of its economics, became a convenient and inexpensive supplier of pure aluminum hydroxide for the chemical industry in addition to its main function of producing metallurgical alumina for the aluminum industry. This situation remains unchanged today.

The various forms of aluminum hydroxides and oxides and their structure and properties have been of interest to chemists since the early 17th century, but the development of their industrial uses goes back only about 50 years. Much pioneering research was done at Alcoa Research Laboratories during this period to find applications for alumina products in the chemical industry. This early work was presented by Francis C. Frary, Director of Alcoa Research Laboratories, in his Perkin Medal (awarded by the Society for Chemical Industry) lecture of 1946 entitled "Adventures with Alumina". The word "adventure" was appropriate in relation to the pioneering nature of this work and the uncertainties connected with the start of a new industry. The industry has since grown and matured, and alumina chemicals have assumed positions of importance in the industrial chemical market. Various alumina products are now basic to manufacturing processes in a score of industries.

The following paragraph from Frary's paper (1) is an eloquent description of the versatility of alumina and an apt introduction to this book.

0065-7719/86/0184/0001$06.00/1
© 1986 American Chemical Society

Aluminum oxide and its hydrates present a variety of amazing contrasts. From the hardness of sapphire to a softness similar to that of talc, from an apparent density of over 200 pounds to one of about 5 pounds per cubic foot, from high insolubility and inertness to ready solubility in acids or alkalis and marked activity, the properties can be varied over wide limits. Some forms flow and filter like sand; others are viscous, thick, unfilterable, or even thixotropic. Crystals may be of any size down to a fraction of a micron in diameter, with various allotropic forms, and there are also amorphous forms. Some varieties have a high adsorptive power, others none at all. Some are catalytically active, others inactive. But they are all converted into α-alumina (corundum) if heated hot enough and long enough.

Alumina in various forms is one of the largest volume, pure inorganic chemicals produced in the world today: production amounts to nearly 40 million metric tons per year. Although production of aluminum metal is the principal consumer of alumina, an increasing amount—8–10% of world production—is finding applications in the chemicals field. Uses include fillers, adsorbents, catalysts, ceramics, abrasives, and refractories.

The principal world source of aluminas is the ore bauxite, a rock consisting chiefly of aluminum hydroxide minerals. Current mining activity is concentrated in Australia, Guinea, Jamaica, Suriname, Yugoslavia, Hungary, Greece, and more recently Brazil. Bauxite itself has some chemical uses but most of it is refined by the Bayer process to remove impurities, mainly oxides of iron, titanium, and silica, to produce a greater than 99% pure alumina product. A small amount of alumina is also produced by the "sinter process", usually from inferior-grade bauxites or nonbauxitic raw materials, but the volume of this production is considerably smaller than that from Bayer operations.

An intermediate step in the refining of bauxite to alumina is the precipitation of aluminum trihydroxide (gibbsite). Aluminum hydroxide is the starting point of numerous alumina chemical products. This close relationship is displayed in Figure 1.1.

Recently, the Ziegler process for the manufacture of linear alcohols, which uses an aluminum intermediate, has become an important source of chemical-grade aluminas. Because of its higher purity, the aluminum monohydroxide (boehmite) obtained as a byproduct of this process is being increasingly used in the preparation of catalysts and catalyst supports and in special ceramic applications.

Some alumina chemical products are also produced by neutralization of aluminum salts and aluminates. These latter chemicals are usually produced from the pure aluminum trihydroxide obtained from the Bayer or sinter process for refining bauxite.

The present-day alumina industry was basically founded for, and is still oriented toward, the supply of metallurgical-grade alumina for the produc-

1. INTRODUCTION

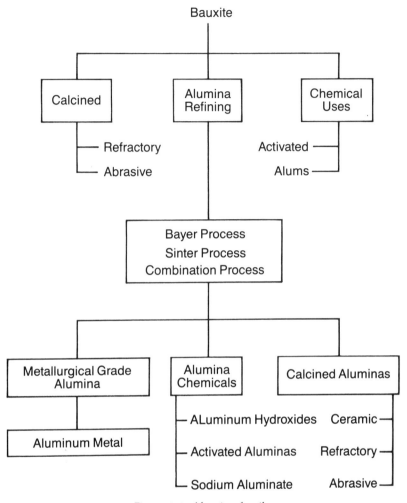

Figure 1.1. Alumina family.

tion of aluminum metal. However, with the development and growth of applications and markets for alumina chemicals, all the major alumina producers have, over the years, converted a part of their capacity to the production of various grades of alumina chemicals. Thus, because alumina chemicals are a part of a stable, large-volume production operation, their supply and cost have remained reliable and stable.

Industrial alumina chemicals can be divided into three broad categories: (1) aluminum hydroxides, (2) activated aluminas, and (3) calcined aluminas. The compound sodium aluminate has been traditionally considered an alumina chemical because of its close relationship to alumina production processes and because it is a ready source of aluminum hydroxide.

Of the above groups, the hydroxides, particularly the trihydroxide obtained from the Bayer process, have the largest volume chemical uses—principally as fillers and as the basic starting material for the preparation of other commercial aluminum compounds. The hydroxides are also the raw materials for activated and calcined alumina products.

Activated aluminas are obtained by controlled thermal dehydration of aluminum hydroxides. The dehydrated product is a porous, high-surface-area material that has found extensive applications in adsorption and catalysis. Appropriate forming technology has been developed to yield products of the desired shape and mechanical strength suitable for industrial applications.

Calcined aluminas, produced by calcination of aluminum hydroxides at higher temperatures, are used in large quantities by the ceramic, refractory, and abrasive industries.

The subject of calcined aluminas and their ceramic uses is of considerable technological importance. But this topic has not been included in this book. The reasons for this exclusion are that traditionally the ceramic industry has always been considered separately from the chemical industry and also that other books that have discussed this subject quite extensively are available (2).

This book begins (Chapter 2) with a classification of aluminum hydroxides and oxides on the basis of their molecular and crystalline structure. The information has a direct bearing on their properties, applications, and thermal behavior. This discussion is followed by chapters (3 and 4) describing their industrial production processes and applications. These applications, developed over the past 50 years, are the result of sustained research cooperation between alumina producers and the user industries. Chemical, physical, and structural changes accompanying thermal dehydration of aluminum hydroxide are examined next (Chapter 5), and this information is followed by a description of production processes for activated alumina products (Chapter 6). Adsorbent and catalytic properties of activated aluminas and their applications are discussed in two subsequent chapters (7 and 8). The next chapter (9) is a description of sodium aluminate as an industrially important alumina chemical.

The last chapter of the book (Chapter 10) discusses some economic aspects of the alumina chemicals industry. Information presented includes the structure of the industry, major producers, type and volume of products, and available cost data.

The book aims at collecting and presenting in an organized way the large amount of information available on industrially important alumina chemicals. At present this information is scattered in diverse technical publications. Systematic examination of this information confirms Frary's description of the versatile nature of alumina.

1. INTRODUCTION 5

Alumina, although now an established chemical, still retains that spark of novelty and unpredictability that makes it an interesting chemical for scientific research. In recent times considerable research effort has been expended to understand its surface structure, particularly after dehydration, and to relate this structure to its adsorptive and catalytic properties. This work has led to investigations of methods of controlling and modifying the surface properties to enhance selectivity. Control of pore volume and size distribution and surface modification are now practical technologies having considerable impact on catalytic and chromatographic adsorption processes employing alumina. This area of application of aluminas is certain to grow with the development of separation processes for the biotechnology industry. Other biotechnology-related applications are the uses of alumina as an inert, porous substrate in biological reactors and immobilized enzyme systems and as a carrier in controlled-release chemical systems.

There have been recent reports (3) of the preparation and characterization of alumina membranes with ultrafine pores for separation uses in water desalination, ultrafiltration processes in the food industry, separation of gas mixtures, and other similar processes. An inert, inorganic material such as alumina carries certain advantages in these types of membrane separation processes such as the ability to withstand high temperatures, aggressive products, high pressures, and a high degree of mechanical abuse.

The ionic conductivity properties of β-alumina have gained prominence following its use in the sodium–sulfur battery. Investigation and possible practical applications of these properties of the β-alumina group of compounds still remain largely unexplored and are bound to become fascinating research subjects in the future.

Literature Cited

1. Frary, F. C. *Ind. Eng. Chem., Anal. Ed.* **1946**, *38*, 129.
2. Gitzen, W. H. *Alumina as a Ceramic Material;* American Ceramic Society: Columbus, OH, 1970.
3. Leenars, A. F. M.; Keizer, K.; Burggraaf, A. J. *J. Mater. Sci.* **1984**, *19*, 1077–1088.

2

Aluminum Oxide and Hydroxides

Historical

The earliest reference to an aluminum compound dates back to the fifth century B.C when Herodotus mentioned "alum". Materials with an astringent taste were named "alumen" by the Romans; their use as a mordant was described by Pliny around A.D. 80. The term "alumina" is likely to have originated from the word alumen (1). Studies of the composition of alum continued in the Middle Ages. Hoffman (2) in 1722 proposed that the base of alum was a true distinct earth, and Pott (3) called it "thonichte erde" or "terre argilleuse" (clay earth). In 1754, Marggraf (4) could show that a distinct compound existed in both alum and clays. De Morveau (5) in 1786 proposed the name "alumine" for the base of alum. Greville (6) described a mineral from India that had the composition Al_2O_3. He named the mineral "corundum", which he believed to be the native name of the stone. The mineral "diaspore" was described by Hauy (7) in 1801 and was analyzed by Vauquelin (8) in 1802 and shown to be $Al_2O_3 \cdot H_2O$. Dewey (9) in 1820 found a mineral he called "gibbsite" in honor of Gibbs, an American mineralogist. Analyses of this mineral by Torrey (10) in 1822 corresponded with the formula $Al(OH)_3$ or $Al_2O_3 \cdot 3H_2O$. The name "hydrargillite" was given to a similar mineral found in the Urals. The latter term is more widely used outside the United States.

The name bauxite was derived from the French province Les Baux and is widely used to describe aluminum ore containing high amounts of aluminum hydroxides. Böhm and Nichassen (11) identified an isomer of diaspore by using an X-ray diffraction method and showed that a purified bauxite from Les Baux consisted of this form. The name böhmit (boehmite) was given to this compound by De Lapparent (12). Böhm also discovered a second modification of $Al(OH)_3$ isomeric with gibbsite (13) that was named bayerite by Fricke (14), who erroneously thought it to be a product of the Bayer process. In the next year Fricke recognized his error when the Bayer product was identified as gibbsite. Van Nordstrand et al. (15) reported a third form of the trihydroxide, which was later named nordstrandite in his honor. This mineral has been reported to occur naturally in several areas.

Rankin and Merwin (16) assigned the prefix β to a high-temperature alumina that, according to later investigations (17), was found to contain alkali or alkaline earth atoms. The pure aluminum oxide corresponding to corundum was differentiated as α-alumina (α-Al_2O_3). Several forms of β-alumina have been identified, and the foreign cation is now included as part of the name, that is, sodium β-aluminum.

Aluminum Hydroxides

The existence of a number of hydroxides of aluminum of varying physical and chemical characteristics has been a key factor in the development of the wide range of industrial alumina chemicals available today.

A general classification of the various modifications of aluminum hydroxides is shown in Figure 2.1. The most well-defined crystalline forms are the three trihydroxides [$Al(OH)_3$]—gibbsite (also called hydrargillite in the European literature), bayerite, and nordstrandite. In addition, two modifications of aluminum oxide–hydroxide ($AlOOH$)—boehmite and diaspore—have been clearly defined. Besides these well-defined crystalline phases, several other forms have been claimed in the literature. These forms are, however, incompletely studied and described. Doubts remain as to whether they are truly new phases or simply forms with distorted lattices containing adsorbed or interlamellar water and impurities.

Gelatinous hydroxides may consist of predominantly X-ray-indifferent (amorphous) aluminum hydroxide or gelatinous boehmite (pseudo-boehmite). The X-ray diffraction pattern of the latter shows broad bands that coincide with strong reflections of the well-crystallized oxide–hydroxide boehmite.

Identification of the different hydroxides is best carried out by X-ray diffraction methods. A compilation of the X-ray diffraction patterns is given by Wefers and Bell (18). Recent measurements made at Alcoa Laboratories are given in Table 2.I (19). Mineralogical and structural data are listed in Tables 2.II and 2.III.

Gibbsite. The crystalline aluminum trihydroxide gibbsite is commonly associated with bauxite deposits of the tropical region. Technically, gibbsite is the most important alumina chemical. Its production is an intermediate stage in the production of alumina (Al_2O_3) from bauxite by the Bayer process (Chapter 3).

The particle size of gibbsite varies from 0.5 to nearly 200 μm depending on the method of preparation. The smaller crystals are composed of hexagonal plates and prisms while the larger particles normally appear as agglomerates of tabular and prismatic crystals (Figure 2.2). The basic crystal habit is pseudohexagonal tabular.

2. ALUMINUM OXIDE AND HYDROXIDES

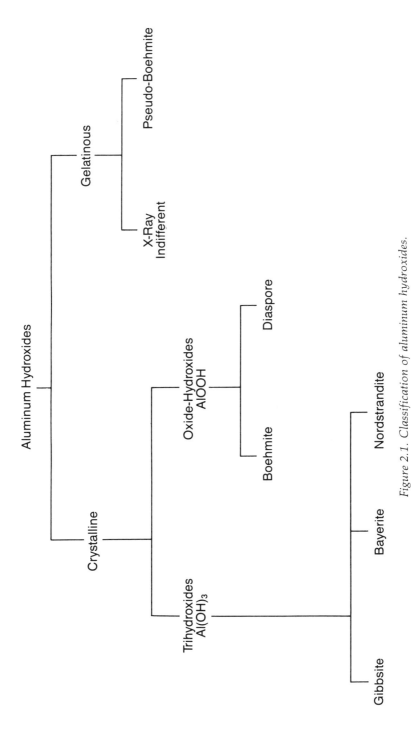

Figure 2.1. Classification of aluminum hydroxides.

Table 2.I.
X-ray Powder Diffraction Patterns for Aluminum Hydroxides and Oxide

Gibbsite		Bayerite		Nordstrandite		Boehmite		Diaspore		Corundum	
d	I	d	I	d	I	d	I	d	I	d	I
4.82	100	4.72	100	4.790	10	6.110	100	4.710	13	3.479	74
4.34	40	4.36	70	4.373	2	3.164	65	3.990	100	2.552	92
4.30	20	3.19	25	4.310	4	2.346	53	3.214	10	2.379	42
3.35	10	3.08	1	4.205	4	1.980	6	2.558	30	2.165	1
3.31	6	2.69	3	4.153	4	1.860	32	2.434	3	2.085	100
3.17	8	2.45	3	3.880	3	1.850	27	2.386	5	1.740	43
3.08	4	2.34	6	3.600	3	1.770	6	2.356	8	1.601	81
2.44	15	2.28	3	3.462	2	1.662	13	2.317	56	1.546	3
2.42	4	2.21	67	3.022	2	1.527	6	2.131	52	1.510	7
2.37	20	2.14	3	2.815	2	1.453	16	2.077	39	1.404	32
2.28	4	2.06	2	2.706	1	1.434	9	1.901	3	1.374	48
2.23	6	1.97	3	2.490	1	1.412	1	1.815	8	1.276	2
2.15	8	1.91	1	2.478	3	1.396	2	1.733	3	1.289	16
2.03	12	1.83	1	2.451	2	1.383	6	1.712	15	1.190	6
1.98	10	1.76	1	2.390	7	1.369	2	1.678	3	1.160	1
1.95	2	1.71	26	2.261	7	1.312	15	1.633	43	1.147	4
1.90	7	1.68	2	2.217	1	1.303	3	1.608	12	1.138	1
1.79	10	1.64	1	2.029	1	1.224	1	1.570	4	1.125	5
1.74	9	1.59	4	2.013	5	1.209	2	1.522	6	1.099	6
1.67	9	1.56	2	1.898	5	1.178	3	1.480	20	1.083	3
1.65	3	1.55	4	1.777	4	1.171	1	1.431	7	1.078	7
1.63	1	1.52	1	1.717	1	1.161	3	1.423	12	1.043	13
1.58	3	1.48	1	1.698	1	1.134	5	1.400	6	1.018	1
1.57	1	1.47	1	1.680	1	1.092	1	1.376	16	0.998	11
1.55	2	1.45	7	1.667	1	1.046	2	1.340	5	0.982	2
				1.647	2						
				1.591	2						
				1.569	2						
				1.545	2						
				1.510	3						
				1.475	3						
				1.438	6						
				1.400	2						

NOTE: The data are interplanar spacings (d) and relative line intensities (I) (filtered Cu Kα radiation).

The usual method of preparation of gibbsite is by crystallization from a caustic aluminate (e.g., $NaAlO_2$) solution either by (1) neutralization by CO_2 or (2) crystallization on seed crystals of gibbsite from a caustic aluminate solution supersaturated with respect to $Al(OH)_3$. These methods are used for the commercial production of the trihydroxide.

$$2NaAlO_2 + CO_2 + 3H_2O \rightarrow 2Al(OH)_3 + Na_2CO_3 \quad (1)$$

$$NaAlO_2 + 2H_2O \rightarrow Al(OH)_3 + NaOH \quad (2)$$

Low-temperature precipitation at less than 40 °C causes heavy nucleation and a fine product. At temperatures above 75 °C, only crystal growth

Table 2.II.
Structures of Aluminum Hydroxides and Oxide

Phase	Formula	Crystal System	Space Group	Molecules per Unit Cell	Unit Axis Length (Å)			Angle
					a	b	c	
Gibbsite	Al(OH)$_3$	monoclinic	C_{2h}^5	4	8.680	5.070	9.720	94°34'
Gibbsite	Al(OH)$_3$	triclinic		16	17.330	10.080	9.730	94°10'
								92°08'
								90°00'
Bayerite	Al(OH)$_3$	monoclinic	C_{2h}^5	2	5.060	8.670	4.710	90°16'
Nordstrandite	Al(OH)$_3$	triclinic	C_1^1	4	8.750	5.070	10.240	109°20'
								97°40'
								88°20'
Boehmite	AlOOH	orthorhombic	D_{2h}^{17}	2	2.868	12.227	3.700	
Diaspore	AlOOH	orthorhombic	D_{2h}^{16}	2	4.396	9.426	2.844	
Corundum	Al$_2$O$_3$	hexagonal (rhombic)	D_{3d}^6	2	4.758		12.991	

SOURCE: Reproduced with permission from Reference 18. Copyright 1972, Aluminum Company of America.

occurs, giving rise to large, well-crystallized aggregates composed of hexagonal rods and prisms (20). Physical properties of gibbsite are given in Table 2.III.

Gibbsite usually contains several tenths of a percent of alkali metal ions; the technical product from the Bayer process contains up to 0.3% Na$_2$O, which cannot be washed out even with dilute HCl. Although several authors (21, 22) maintain that these alkali ions are an essential component of gibbsite structure, this statement has not been established unequivocally. Wefers (21) and Misra and White (20) have reported on the effect of sodium and potassium ions on the morphology of gibbsite. Crystals grown from sodium aluminate solution had a tabular habit while pseudohexagonal elongated prisms prevailed when gibbsite was precipitated from potassium aluminate solutions.

A possible structure for gibbsite was suggested by Pauling (23) and was later confirmed by Megaw (24) and refined by Saalfeld (25). The basic structural unit of gibbsite is a layer of Al ions sandwiched between two sheets of closely packed hydroxyl ions. The Al ions occupy two-thirds of the octahedral interstices within the layers in contrast to the brucite [Mg(OH)$_2$] structure where all the sites are occupied. Each double layer is positioned with respect to its upper and lower neighbors in such a way that, ideally, hydroxyl ions of adjacent planes are directly opposite each other (Figure 2.3). Thus, the sequence of OH ions in the direction perpendicular to the planes is AB–BA–AB–BA. Hydrogen bonds operate between OH groups of adjacent double layers. Proton magnetic resonance studies by Glemser (26) and Kroon and Stolpe (27) have explained the role of these hydrogen bonds in the structure of gibbsite.

Table 2.III.
Physical Properties of Aluminum Hydroxides and Oxide

Compound	Formula	Color and Luster	Index of Refraction[a]			Moh Hardness	Density (g/cm³)	wt % of Al_2O_3	wt % of H_2O	$\Delta H^{F,b}$ (Kcal/mol)	sp heat (cal/g)[c]
Gibbsite	$Al(OH)_3$, $Al_2O_3 \cdot 3H_2O$	white, pearly, vitreous	1.568	1.568	1.587	2.5–3.5	2.42	65.4	34.6	−309.1	0.282[d]
Bayerite	$Al(OH)_3$, $Al_2O_3 \cdot 3H_2O$	white					2.53	65.4	34.6	−307.9	
Nordstrandite	$Al(OH)_3$, $Al_2O_3 \cdot 3H_2O$	white						65.4	34.6		
Boehmite	$AlOOH$, $Al_2O_3 \cdot H_2O$	white	1.649	1.659	1.665	3.5–4	3.01	85.0	15.0	−236.7	0.261[e]
Diaspore	$AlOOH$, $Al_2O_3 \cdot H_2O$	white, brilliant, pearly	1.702	1.722	1.750	6.5–7	3.44	85.0	15.0	−238.9	0.211[f]
Corundum	Al_2O_3	white, pearly adamantine				9	3.98	100.0	0.0	−400.4	0.183[g]

SOURCE: Reproduced with permission from Reference 18. Copyright 1972, Aluminum Company of America.
[a]The average index of refraction for bayerite is 1.583. Index of refraction values for corundum are as follows: ϵ, 1.7604; ω, 1.7686.
[b]Heat of formation at 298.16 K.
[c]Specific heat at 20 °C.
[d]Specific heat of gibbsite (cal/g) = $0.2694 + 6.43 \times 10^{-4} \times t$ (t is in degrees Celsius).
[e]Specific heat of boehmite (cal/g) = $0.2598 + 7 \times 10^{-5} \times t$ (t is in degrees Celsius).
[f]Specific heat of diaspore (cal/g) = $0.2085 + 1.27 \times 10^{-4} \times t$ (t is in degrees Celsius).
[g]Specific heat of corundum (cal/g) = $0.34833 - 8.02 \times 10^{-6} \times (273.1 + t) - 47.852/(273.1 + t)$ (t is in degrees Celsius).

2. ALUMINUM OXIDE AND HYDROXIDES

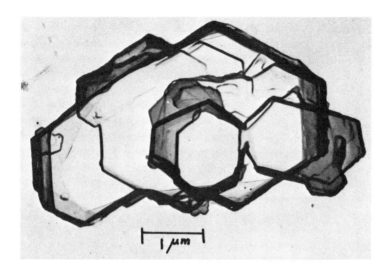

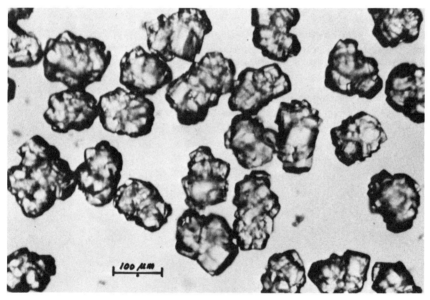

Figure 2.2. Gibbsite crystals.

The structure of gibbsite is in fact somewhat distorted from the ideal; the Al–O octahedra are not regular but have their shared edges longer than the remainder. The layers are somewhat displaced by one another in the direction of the a axis; this displacement results in a monoclinic cell. In gibbsite the two layers in each cell have dimensions $a = 8.64$ Å, $b = 5.07$ Å, $c = 9.72$ Å, $94°34'$, and the space group $P2_1/n$ or C_{2h}^5 and contain eight Al(OH)$_3$. Saalfeld (25) reported triclinic symmetry in larger gibbsite single

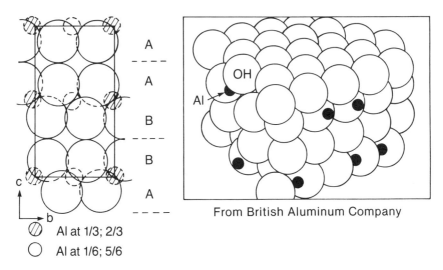

⊘ Al at 1/3; 2/3
○ Al at 1/6; 5/6

Figure 2.3. Gibbsite structure.

crystals from the Urals. Saalfeld explained this result by assuming displacement of the layers in the b axis as well.

According to Wefers (28), the alkali ions are essential to the stabilization of the gibbsite structure and probably occupy vacancies in the hydroxyl ion packing. The charge neutrality within the octahedral layers is thereby disturbed, the excess positive charges causing stronger bonding between the layers. The incorporation of potassium ions (the radius of which is nearly the same as that of the OH ion) leads to the formation of elongated, morphologically perfect pseudohexagonal prisms. On the other hand, the incorporation of the much smaller sodium ions results in the formation of less well-developed crystals with mainly tabular habit; these crystals often exhibit marked spiral growth.

Bayerite. Except for rare occurrences, bayerite is not found in nature. It has been prepared synthetically by several methods:

1. A simple method for the preparation of crystallographically pure bayerite is described by Schmäh (29). In this procedure, amalgamated aluminum is reacted with water at room temperature. The product is a crystalline powder (Figure 2.4a) with a well-defined X-ray diffraction pattern.

2. A second method is neutralization of aluminum salts with NH_4OH followed by aging at room temperature.

3. Another method is bubbling CO_2 through aluminate solutions at room temperature (30). The appearance of this product is shown in Figure 2.4b.

4. Spontaneous precipitation on bayerite seed from a highly supersaturated aluminate solution is a method.

5. A fifth method is rehydration of ρ- (transition) alumina.

The product from methods 2–5 is rarely pure bayerite. This product usually contains appreciable amounts of other aluminum hydroxides.

(a)
× 2400

(b)
× 2400

Figure 2.4. (a) Schmäh bayerite. (b) Bayerite produced by CO_2 gassing of sodium aluminate liquor.

Unlike gibbsite, bayerite can be prepared without any alkali ions, so no controversy exists about possible stabilization of its structure by these ions. Uncertainties regarding the structure and stability field of bayerite do exist, however. Chistiyakova (31) maintains that bayerite is stable below 20 °C in alkali aluminate solutions. Other works (21, 22) give evidence that bayerite is only stable in the Al_2O_3–H_2O binary system and converts irreversibly to gibbsite in the presence of alkali (Na and K) metal ions. Reports of structural discrepancies arise from the fact that bayerite does not form well-defined single crystals that are large enough for proper structural analysis. According to Wefers (21, 28), bayerite particles occur as "somatoids". These particles are of uniform shape but are not enclosed by well-defined crystal faces. These somatoids are formed by the loose stacking of $Al(OH)_3$ layers in a direction perpendicular to the basal plane. As with gibbsite, the bayerite lattice is also composed of double layers of OH ions. The layers are, however, considered to be arranged in the AB–AB–AB sequence; hydroxyl ions of the third layer lie in the depression between the OH positions of the second. This approximately hexagonal close packing explains the higher density of bayerite compared to that of gibbsite. The crystal class and space group of bayerite have not yet been clearly established. In an earlier investigation by Montoro (32), the cell is described as trigonal with $a = 5.01$ Å, $c = 4.76$ Å, and space group D_{3d} or $H\bar{3}M$. Yamaguchi and Sakamoto (33) also had the same results. Lippens (34) found bayerite to be orthorhombic, while Bezjak and Jelenic (35) calculated the triclinic space group $P1$.

Physical properties of bayerite are given in Table 2.III. Bayerite is a commercially available technical product. An example is Alcoa C-37 aluminum hydroxide.

Nordstrandite. Van Nordstrand et al. (15) first reported this form of aluminum trihydroxide in 1956. Since then, nordstrandite has been discovered to occur naturally in several countries. Many authors have reported nordstrandite in mixtures with other hydroxides. Hauschild (36) has shown that pure nordstrandite can be prepared by reacting aluminum, aluminum hydroxide gel, or hydrolyzable aluminum compounds with aqueous solutions of alkylene diamines, especially ethylenediamine.

Van Nordstrand and co-workers originally considered this form to be a screw dislocation polymorph of the gibbsite lattice. As proposed by Lippens (34), the nordstrandite structure is also assumed to consist of double layers of hydroxyl ions with aluminum in two-thirds of the octahedral holes. The stacking of the double layers is proposed to be AB–AB–BA–BA, which is a combination of the stacking found in gibbsite and bayerite. Saalfeld and Mehrotra (37) have analyzed the structure of the natural single crystal of nordstrandite and suggest a triclinic unit cell.

Available data on physical properties of nordstrandite are given in Table 2.III. Although technical production of nordstrandite is covered by several patents (38), no reports of commercial production or use can be found.

2. ALUMINUM OXIDE AND HYDROXIDES

Aluminum Oxide-Hydroxide (AlOOH)

Boehmite. The aluminum oxide-hydroxide boehmite is a major constituent of bauxite deposits of Europe. Boehmite is also found in association with gibbsite in some tropical bauxites in Australia, Asia, and Africa.

Well-crystallized boehmite is prepared from the trihydroxides by hydrothermal transformation under water at temperatures above 150 °C. Figure 2.5 shows boehmite crystals of 5–10-µm size produced by this method. Higher temperatures and the presence of alkali increase the rate of transformation (39). Fibrous (acicular) boehmite is obtained under acidic conditions (40). Hydrothermally produced boehmite usually contains water in excess (1–2%) of the 15% stoichiometric water content of boehmite.

The structure of boehmite (Figure 2.6) consists of double layers in which the oxygen atoms are in cubic packing. Hydroxyl ions of one double layer are located over the depression between OH ions in the adjacent layer. Double layers are linked by hydrogen bonds between hydroxyl ions in neighboring planes.

Physical properties of boehmite are given in Table 2.III. There is some technical production and use of hydrothermally produced boehmite.

Diaspore. The aluminum oxide–hydroxide diaspore is a major constituent of bauxites of Greece, European Russia, and China. In the United States, diaspore has been found in clays in Missouri and Pennsylvania.

Figure 2.5. Hydrothermal boehmite (9,000 × magnification).

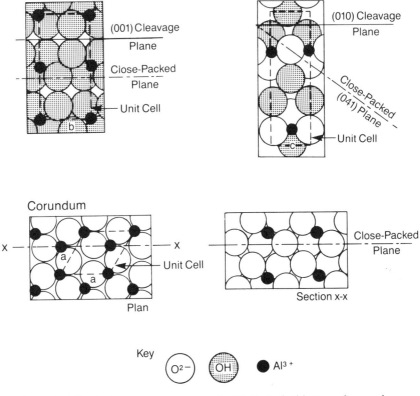

Figure 2.6. Boehmite structure compared with that of gibbsite and corundum. Reproduced from Reference 64. Copyright 1969 BA Chemicals Ltd.

Laubengayer and Weisz (41), Ervin and Osborn (42), Roy and Osborn (43), Kennedy (44), and Neuhaus and Heide (45) have shown that hydrothermal conversion of all aluminum hydroxides between 225 and 400 °C and pressures above 150 bar results in the formation of diaspore when seed crystals of diaspore are present. However, diaspore crystals can be grown at temperatures below 120 °C epitaxially on goethite (α-FeOOH), which is structurally isomorphous with diaspore (46).

The oxygen atoms are nearly equivalent in the diaspore structure, each being joined to another oxygen by way of a hydrogen ion and arranged in hexagonal close packing. This compact arrangement accounts for the greater density and stability of diaspore.

Physical property data for diaspore are given in Table 2.III. Because of the high temperature and pressure conditions required for its preparation, diaspore has not been produced or used commercially.

Gelatinous Aluminum Hydroxides

Apart from the crystalline forms, aluminum hydroxide also occurs in the amorphous and gelatinous forms. The composition and properties of these gels depend largely on the method of preparation. These gel products are of considerable technical and commercial importance.

The solubility of $Al(OH)_3$ at different pH values is shown in Figure 2.7. The solubility is very low in the pH range of 5-8. The amphoteric behavior of the hydroxide is evident in this diagram. $Al(OH)_3$ dissolves readily in strong acids and bases. In the pH range of 4-9, a slight variation in pH toward the neutral value can cause rapid and voluminous precipitation of colloidal hydroxide. Being hydrophilic, the colloidal precipitate readily forms gels.

Aluminum hydroxide gels formed by neutralization from either acid or basic solutions contain considerable excess water and variable amounts of anions. Even after prolonged drying at 100-110 °C, the water content of gels can be as high as 5 mol of H_2O/mol of Al_2O_3. The water content, size of primary particle of the precipitate, and specific surface area vary with the conditions of precipitation.

Studies of gelatinous aluminum hydroxides have generally followed three methods of preparation:

1. neutralization, either of aluminum salts with alkali or of alkali aluminates with CO_2 or acids

2. decomposition (mostly by hydrolysis) of aluminum organic compounds such as aluminum methylate, ethylate, alcoholates, isopropoxide, butoxide, etc.

3. reaction of water with aluminum metal (usually activated by amalgamation)

Because of their technical importance, the preparation and properties of the so-called alumina gels have been widely studied. An extensive review is given by Wefers and Bell (*18*). They conclude that the various methods of preparation lead to three types of gelatinous products. The predominant solid phases in these gels can be (1) X-ray indifferent (amorphous); (2) gelatinous boehmite, also termed pseudo-boehmite; and (3) finely crystallized trihydroxide (i.e., bayerite, nordstrandite, or gibbsite).

Except for material prepared at a pH above 7, or at elevated temperatures, the initial product is usually always X-ray indifferent. Transformation to crystalline hydroxide (aging) occurs; the rate is dependent on the OH ion concentration and temperature. The first X-ray crystalline phase in the aging sequence is similar to that of boehmite and is called gelatinous or pseudo-boehmite. The diffraction pattern of gelatinous boehmite shows broad

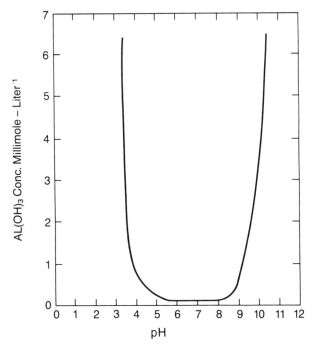

Figure 2.7. Solubility of Al(OH)$_3$ as a function of pH.

bands that coincide with the strong reflections of the well-crystallized oxide–hydroxide boehmite. This similarity has been the cause of considerable confusion and has resulted in extensive investigations into the structure and characteristics of pseudo-boehmite.

Figure 2.8 (22) shows the X-ray diffraction patterns of an aluminum hydroxide precipitated at 20 °C and at a pH of 11 from an aluminum chloride solution. Figure 2.8 shows the gradual development of the crystalline order of the hydroxide. The factors influencing the degree of order include the residence time of the precipitated product in the mother liquor, the pH values, and the ions present in the solution. Pseudo-boehmite is the first product of aging and is the precursor stage of the trihydroxide. Diffraction diagrams of this aging product are shown in lines A and B of Figure 2.8. The peaks of the broad reflexes are in good agreement with the position of strong boehmite lines. As aging proceeds, the final outcome of the ordering process is the formation of bayerite.

Lippens (34) was able to demonstrate that the reflexes of pseudo-boehmite are broadened considerably not only because of the small particle size but also because of variable distances of the AlOOH double chains that form the basic structural element of pseudo-boehmite as well as of well-crystallized boehmite. The 020 line (b crystallographic dimension) increases from

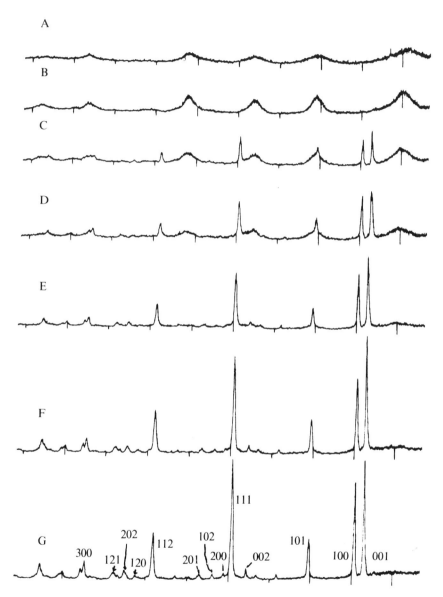

Figure 2.8. Increasing degree of order of freshly precipitated Al(OH)$_3$ (A) and after (B) 4, (C) 8, (D) 12, (E) 16, (F) 20, and (G) 24 h aging under the mother liquor. Reproduced with permission from Reference 47a. Copyright 1971 Ferdinand Enke Verlag.

about 6.15 Å in well-crystallized boehmite to as much as 6.6–6.7 Å in pseudo-boehmite. The increase in the b spacing is accompanied by an increase in the water content over the stoichiometric formula for boehmite ($Al_2O_3 \cdot H_2O$). Papee et al. (47b) postulated that the excess water is located between the boehmite-like layers as molecular water; this water enlarges the b dimension of the crystal structure. Lippens (34) reported as much as 30.7% water in pseudo-boehmite compared to the stoichiometric 15% for the boehmite formula. After correcting for the effect of small angle scattering on the X-ray diffraction pattern, Lippens found a linear relationship between the increased b dimension and the mole fraction of excess water: an increase of 1.17 Å for each mole of excess H_2O. The above interpretation has been questioned by Baker and Pearson (48), who have used X-ray diffraction data and nuclear magnetic resonance spectra to propose an alternative model for the structure of pseudo-boehmite that accounts for the excess water. Baker and Pearson suggest that pseudo-boehmite can be considered to be "ultramicrocrystalline boehmite", where all the excess water is simply the terminal water molecules attached to the ends of the relatively small boehmite chain structure of the crystallites.

In contact with a mother liquor having a pH greater than 7, gelatinous or pseudo-boehmite transforms into crystalline trihydroxides. In the absence of alkali ions and in the presence of ammonia, amines, etc., the transformation produced is bayerite (or nordstrandite); conversion to gibbsite occurs in the presence of alkali metal ions. The rate of transformation increases with pH and temperature. The transformation sequence may be represented by the scheme proposed by Ginsberg et al. (22):

$$Al^{3+} + 3OH^- \xrightarrow{pH\ 8} \text{basic Al salt} \xrightarrow{pH\ 8}$$

$$\text{X-ray amorphous gel} \xrightarrow{pH\ 9} \text{gelatinous boehmite} \xrightarrow{pH\ 10}$$

$$\text{bayerite or nordstrandite} \xrightarrow{+Na^+ \text{ or } K^+} \text{gibbsite}$$

Aluminum Oxide (Corundum)

Corundum (Al_2O_3) is the only stable oxide of aluminum and is the final product of thermal and hydrothermal decomposition of all aluminum hydroxides. Natural corundum is typically found in metamorphic and igneous rocks. Corundum of lower purity is used as an abrasive; it is the chief component of emery deposits. The colored varieties of corundum suitable for gemstones are known as ruby for the red variety and sapphire for the blue variety.

The properties of corundum are given in Tables 2.II and 2.III. The crystal structure of corundum was fully investigated by Pauling and Hendricks (49) following earlier work by Bragg and Bragg (50). Numerous other investigations in later years have confirmed this structure. The lattice of corundum is composed of hexagonally closest packed oxygen ions forming layers parallel to the (0001) plane. Only two-thirds of the octahedral interstices are occupied by aluminum ions. The lattice may be described roughly as an arrangement of alternating layers of oxygen and aluminum ions.

β-Alumina and Related Compounds

A new alumina phase, believed to be a polymorph of α-Al_2O_3 (corundum), and hence termed β-Al_2O_3, was reported by Rankin and Merwin (16) in 1916. Later investigations by Stillwell (17), Bragg et al. (51), and Beevers and Ross (52) showed that sodium oxide is necessary for the formation of β-alumina. The formula $Na_2O \cdot 11Al_2O_3$ was given to this compound although its composition is variable.

The so-called β-aluminas have now come to represent a group of aluminates having the same, or very similar, structures but variable chemical composition. The 1:11 ($Na_2O:Al_2O_3$) β-Al_2O_3 crystallizes from melts containing aluminum oxide and sodium oxide or other sodium compounds. A 1:5 β-alumina has been prepared by heating α-Al_2O_3 with $NaAlO_2$ or Na_2CO_3 at about 1100 °C (53). Schloder and Mansmann (54) claimed that the β-aluminas containing alkali oxides have the same composition as the 1:6 aluminates of the alkaline earth oxides (e.g., $GaO \cdot 6Al_2O_3$, $BaO \cdot 6Al_2O_3$, and $SrO \cdot 6Al_2O_3$). Bragg et al. (51) reported that the magnesium aluminate is a 1:11 compound.

Beevers and Ross (52) describe the structure of β-alumina as consisting of blocks of cubic closely packed oxygen atoms of the thickness of four closely packed layers. The oxygens of the blocks are held together by aluminum atoms in positions identified with those of Al and Mg in the spinel structure. Adjacent blocks are linked by a layer of oxygen and alkali metal atoms that is not closely packed. The predominant crystal habit of the β-aluminas is that of the hexagonal prism; very thin platelets are common. Properties of β-aluminas are given in Table 2.IV.

Recent interest in β-aluminas is related to its use as a solid electrolyte in the sodium–sulfur secondary battery for storage of electrical energy. The structure of β-alumina results in it being a Na^+ ionic conductor, the Na^+ ions being able to move freely under the application of an electric field. The sodium–sulfur battery, first announced by the Ford Motor Company in 1967 (55), consists of two liquid electrodes, molten sodium and molten sulfur–sodium polysulfide melt, separated by a solid β-alumina electrolyte. When an external circuit is completed, Na^+ ions diffuse through the electrolyte, producing sodium polysulfides as the discharge product, and electrons pass

Table 2.IV.
Structural and Other Properties of β-Aluminas

Phase	Formula	Crystal System	Space Group	Molecules per Unit Cell	Unit Axis Length (Å) a	Unit Axis Length (Å) c	Density (g cm^{-3})	Index of Refraction ε	Index of Refraction ω	Index of Refraction αυ
Sodium β	Na$_2$O·11Al$_2$O$_3$	hexagonal	D_{6h}^4	1	5.58	22.45	3.24	1.635-1.650	1.676	
Sodium β	Na$_2$O·5Al$_2$O$_3$	hexagonal			5.61	33.95	3.30			
Potassium β	K$_2$O·11Al$_2$O$_3$	hexagonal	D_{6h}^4	1	5.58	22.67				
Magnesium β	MgO·11Al$_2$O$_3$	hexagonal	D_{6h}^4	1	5.56	22.55		1.629	1.665-1.680	
Calcium β	CaO·6Al$_2$O$_3$	hexagonal	D_{6h}^4	2	5.54	21.83				
Strontium β	SrO·6Al$_2$O$_3$	hexagonal	D_{6h}^4	2	5.56	21.95				
Barium β	BaO·6Al$_2$O$_3$	hexagonal	D_{6h}^4	2	5.58	22.67	3.69	1.694	1.702	
Lithium ζ	Li$_2$O·5Al$_2$O$_3$	cubic	O_h^7	2	7.90		3.61			1.735

SOURCE: Reproduced with permission from Reference 18. Copyright 1972, Aluminum Company of America.

around the external circuit, producing an open circuit voltage of 2.08 V. The reverse reaction occurs during charging of the battery. A very extensive review of the structure, properties, and production of β-alumina has recently been made by Stevens and Binner (56).

Phase Relationships in the Al_2O_3–H_2O System

The conversion of amorphous precipitated aluminum hydroxide to crystalline forms has been discussed previously under Gelatinous Aluminum Hydroxides. In the absence of alkali metal ions, the final product of transformation is bayerite. Bayerite is relatively stable at 20 °C but converts rapidly to gibbsite at higher temperatures in the presence of Na^+ or K^+ ions. However, the theory that these ions are necessary for the gibbsite structure still remains controversial.

Under the equilibrium vapor pressure of water, crystalline $Al(OH)_3$ converts to AlOOH at about 100 °C. The conversion temperature appears to be the same for all three forms of $Al(OH)_3$. Below 280–300 °C, boehmite is the prevailing AlOOH modification unless the diaspore seed is present. Spontaneous nucleation of diaspore requires temperatures in excess of 300 °C and pressures higher than 200 bar. For this reason, diaspore was considered the high-temperature polymorph of AlOOH in older literature.

The first study of phase transitions in the Al_2O_3–H_2O system was reported by Laubengayer and Weiss (41) in 1943. These workers determined the gibbsite–boehmite conversion temperature to be 155 °C. Boehmite transformed to diaspore above 280 °C; diaspore converted to corundum (α-Al_2O_3) at 450 °C. Ervin and Osborn (42) reported similar results and presented results of their hydrothermal studies as a phase diagram of the Al_2O_3–H_2O system (Figure 2.9). In 1953, Newsome (57) obtained a patent for a method of converting crystalline aluminum hydroxides to α-alumina (corundum) by heating the hydroxides to a temperature of 400–550 °C in the presence of steam at a pressure of 25–176 bar.

The Al_2O_3–H_2O system was reinvestigated by Kennedy (44). Although generally confirming the results of Ervin and Osborn, Kennedy showed that the shape of the boehmite–diaspore phase boundary in the Ervin–Osborn diagram is thermodynamically inconsistent for a solid–solid transition. The $\Delta S/\Delta V$ ratio, which equals the slope dp/dt, normally remains constant over a wide temperature and pressure range for a solid–solid transition. An 8.5% reduction in atomic volume between boehmite and diaspore occurs and should be accompanied by a rather large positive entropy change. These two facts suggest that the boundary separating boehmite and diaspore on the p–t diagram should be reasonably close to a straight line and have a positive slope. However, the phase diagram of Ervin and Osborn shows a boundary having essentially zero slope in the temperature range of 300–400 °C that changes to infinity at a temperature around 270 °C. The $\Delta S/\Delta V$

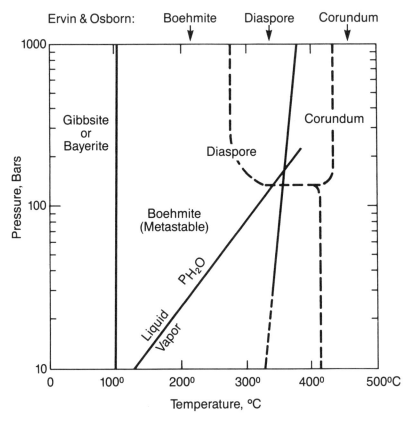

Figure 2.9. The system Al_2O_3–H_2O: (---) Kennedy; (—) Neuhaus and Heide; (–––) Ervin and Osborn.

value changes from zero to infinity within an interval of only 50 °C. Kennedy concluded from this peculiar course of the phase boundary that this diagram does not represent true equilibrium conditions. Probably, as suggested earlier by Laubengayer and Weiss, a rate process rather than an equilibrium boundary was involved. Boehmite is metastable in the Al_2O_3–H_2O system; diaspore is the stable phase.

Neuhaus and Heide (45) confirmed Kennedy's results. In their experiments, diaspore seed crystals were shown to grow at the expense of boehmite at temperatures above 180 °C. Neuhaus and Heide suggested that even though boehmite is metastable, its formation is kinetically favored at lower temperatures and pressures. One reason is that the nucleation energy is lower for boehmite than for the considerably more dense diaspore. In addition, gibbsite and boehmite have similar structures; this fact favors epitaxial growth of boehmite on the trihydroxide. No such structural analogy between diaspore on one hand and boehmite or bayerite on the other exists. Spontaneous diaspore formation was observed by Neuhaus and Heide, in agreement with other workers, from only 300 °C upward and pressures of several hundred bar. At 360 °C and 200 bar, diaspore converted reversibly to corundum. Wefers (46) has shown that in the system $Al_2O_3-Fe_2O_3-H_2O$, the presence of isostructural FeOOH (goethite) lowered the nucleation energy for diaspore so that this AlOOH modification crystallized at temperatures near 100 °C. This observation explains the occurrence of diaspore in clays and bauxite deposits that had never been subjected to high temperatures and pressures. A phase diagram based on the results of Kennedy and Neuhaus and Heide is shown in Figure 2.9.

Torkar and Krischner (58) have reported the formation of several nonequilibrium forms of aluminum oxides in the $Al_2O_3-H_2O$ system in the temperature region of 300–500 °C and low water vapor pressures of 10–200 bar. In hydrothermal studies carried out at Alcoa Laboratories (59), the KI–Al_2O_3 form originally reported by these workers was prepared in a pure form. The material crystallizes in small (approximately 0.5 μm) thin hexagonal plates with a density of 3.66 g/cm^3 and refractive index of 1.73. This form has a water content of about 2.74%, which corresponds to the formula $1.6H_2O \cdot 11Al_2O_3$. The material loses water above 700 °C as it converts to α-Al_2O_3. The product is highly crystalline as shown by the well-defined X-ray diffraction spectrum, which is identical to that reported by Torkar and Krischner. The product "tohdite" reported by Yamaguchi et al. (60) has been shown to be identical with the KI–Al_2O_3 of Torkar and Krischner. A schematic representation of the different products obtained in the studies of Torkar and Krischner (61) at different temperatures and pressures is displayed in Figure 2.10.

Nomenclature

Although the phase fields and structures of the crystalline phases in the system $Al_2O_3-H_2O$ are clearly defined, the nomenclature is still rather unsystematic.

Bayerite, gibbsite (or hydrargillite), and nordstrandite are aluminum trihydroxides and not oxide hydrates. The commonly used term aluminum oxide monohydrate for boehmite and diaspore is also incorrect. Both are true oxide–hydroxides. So far, molecular water has been determined in pseudo-boehmite only. One would not want to consider this material a defined oxide-hydroxide hydrate either, because a pure pseudo-boehmite has been impossible to prepare thus far. The reason is because only short-range ordered zones develop during the topochemical conversion of the X-ray-indifferent, primary solid precipitate. These zones are embedded in extensive areas of the amorphous gel. Increasing the degree of order always im-

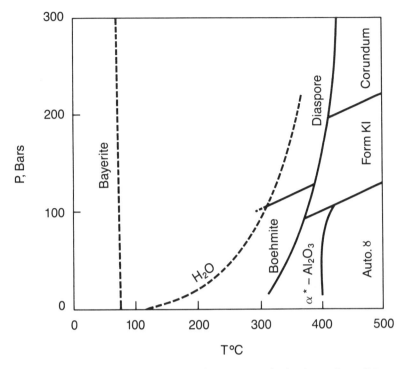

Figure 2.10. Aluminas formed under low-pressure hydrothermal conditions. Reproduced with permission from Reference 58. Copyright 1960, Springer.

plies conversion to crystalline Al(OH)$_3$. The so-called aluminum oxide dihydrate, which was reported by Pentscheff and Zaprianova (62), is also probably a pseudo-boehmite. According to these authors, Al$_2$O$_3$·2H$_2$O has a defined chemical composition but has no specific diffraction diagram. Only the lines of boehmite appeared in the X-ray diffraction spectrum.

Saalfeld et al. (63) were able to remove the sodium ions from the lattice of β-alumina (Na$_2$O·11Al$_2$O$_3$) by leaching with H$_2$SO$_4$ and replace them with H$_2$O molecules. This case is, to date, the only one in which the existence and structure of an alumina hydrate were confirmed. As an intercalation compound, this hydrate has, however, no phase field in the system Al$_2$O$_3$–H$_2$O.

Several nomenclature systems using Greek letters are found in the literature. These systems also lack uniformity and have been a source of confusion. For example, the trihydroxide gibbsite and the oxide–hydroxide boehmite have both been designated as the α forms. However, according to general usage in crystallography, the most densely packed structures—bayerite, diaspore, and corundum—should be "logically" designated as the modifications. In similar fashion, the compounds with a cubic packing sequence, gibbsite and boehmite, should be designated by the symbol γ and nordstrandite as β-Al(OH)$_3$.

Literature Cited

1. Beckman, J. *A History of Inventions, Discoveries and Origins*, 4th ed.; Johnston, W., Trans. (revised and enlarged by W. Francis and J. W. Griffith); Henry G. Bohn: London, 1799.
2. Hoffman, F. *Observationum Physico-Chymicarum Selectorium;* Halae, 1722.
3. Pott, J. H. *Lithogeognosia;* Potsdam, 1746.
4. Marggraf, A. S. *Mem. Acad. Berlin* **1754**, 41.
5. De Morveau, L. B. *Encyclopedia Methodique;* Gyton: Paris, 1786.
6. Greville, C. In *Encyclopaedia Brittanica*, 11th ed.; 1799; Vol. 7, p 207.
7. Hauy, R. J. *Traite Mineral.* **1801**, *4*, 358.
8. Vauquelin, L. N. *Ann. Chim. (Paris)* **1802**, *1*, 42, 113.
9. Dewey, C. *Am. J. Sci.* **1820**, *1, 2*, 249.
10. Torrey, J. *Edinburgh Phil. J.* **1822**, *7*, 388.
11. Bohm, J.; Nichassen, H. *Z. Anorg. Allg. Chem.* **1924**, *132*, 1-9.
12. De Lapparent, J. *C. R. Hebd. Seances Acad. Sci.* **1927**, *184*, 1661-1662.
13. Bohm, J. *Z. Anorg. Allg. Chem.* **1925**, *149*, 203-216.
14. Fricke, R. *Z. Anorg. Allg. Chem.* **1928**, *175*, 249-256.
15. Van Nordstrand, R. A.; Hettinger, W. P.; Keith, C. D. *Nature (London)* **1956**, *177*, 713-714.
16. Rankin, G. A.; Merwin, H. E. *J. Am. Chem. Soc.* **1916**, *38*, 568-588.
17. Stillwell, C. W. *J. Phys. Chem.* **1926**, *30* (11), 1441.
18. Wefers, K.; Bell, G. M. *Oxides and Hydroxides of Aluminum;* Technical Paper No. 19; Aluminum Company of America: Alcoa Center, PA, 1972.
19. Evans, W. T.; Burns, R. A. "Characterization of Alumina Phases"; Alcoa internal report; Aluminum Company of America: Alcoa Center, PA, 1979.
20. Misra, C.; White, E. T. *J. Cryst. Growth* **1971**, *8*, 172-178.
21. Wefers, K. *Naturwissenshaften* **1962**, *49*, 204-205.
22. Ginsberg, H.; Huttig, W.; Stiehl, H. *Z. Anorg. Allg. Chem.* **1962**, *318*, 238-256.
23. Pauling, L. *Proc. Natl. Acad. Sci. U.S.A.* **1930**, *16*, 123.
24. Megaw, H. D. *Z. Kristallogr.* **1934**, *A87*, 185-204.
25. Saalfeld, H. *Neues Jahrb. Mineral., Abh.* **1960**, *95*, 1-87.
26. Glemser, O. *Nature (London)* **1959**, *183*, 943.
27. Kroon, D. J.; van der Stolpe, C. *Nature (London)* **1959**, *183*, 944.
28. Wefers, K. *Metall (Berlin)* **1967**, *21*, 423-431.
29. Schmäh, H. *Z. Naturforsch.* **1946**, *1*, 323-324.
30. Sato, T.; Yamashita, T.; Ozawa, F. *Z. Anorg. Allg. Chem.* **1969**, *370*, 202-208.
31. Chistiyakova, A. A. *Tsvetn. Met.* **1964**, *37* (9), 54.
32. Montoro, V. *Ric. Sci.* **1942**, *13*, 565-571.
33. Yamaguchi, G.; Sakamoto, K. *Bull. Chem. Soc. Jpn.* **1958**, *31*, 140.
34. Lippens, B. C. Ph.D. Thesis, Delft University, 1961.
35. Bezjak, A.; Jelenic, I. Symp. Sur Les Bauxites II, ISCOBA, Zagreb **1963**, 105-111.
36. Hauschild, U. *Z. Anorg. Allg. Chem.* **1963**, *324*, 15-20.
37. Saalfeld, H.; Mehrotra, B. *Ber. Dtsch. Keram. Ges.* **1965**, *42*, 161-166.
38. Hauschild, U. Br. Patent 950 165, 1964.
39. Ginsberg, H.; Koster, M. *Z. Anorg. Allg. Chem.* **1952**, *271*, 41-48.
40. Bugosh, J. U.S. Patent 2 915 475, Dec 1, 1959.
41. Laubengayer, A. W.; Weiss, R. S. *J. Am. Chem. Soc.* **1943**, *65*, 247-250.
42. Ervin, G.; Osborn, E. F. *J. Geol.* **1951**, *59*, 381-394.
43. Roy, R.; Osborn, E. F. *Am. Mineral.* **1954**, *39*, 853-885.

44. Kennedy, G. C. *Am. J. Sci.* **1959**, *257*, 563-573.
45. Neuhaus, A.; Heide, H. *Ber. Dtsch. Keram. Ges.* **1965**, *42*, 167-184.
46. Wefers, K. *Z. Erzbergbau. Mettallhuettenwesen.* **1967**, *20*, 13-9, 71-75.
47a. Ginsberg, H.; Wefers, K. *Aluminum and Magnesium*; Ferdinand Enke Verlag: Stuttgart, 1971.
47b. Papee, D.; Tertian, R.; Biais, R. *Bull. Soc. Chim. Fr.* **1958**, 1301-1310.
48. Baker, B. R.; Pearson, R. M. *J. Catal.* **1974**, *33*, 265-278.
49. Pauling, L.; Hendricks, S. B. *J. Am. Chem. Soc.* **1925**, *47*, 781.
50. Bragg, W. H.; Bragg, W. L. *X-rays and Crystal Structure*; Bell and Sons: London, 1916; p 169.
51. Bragg, W. L.; Gottfried, C.; West, J. *Z. Kristallogr.* **1931**, *77 (3-4)*, 255.
52. Beevers, C. A.; Ross, M. A. S. *Z. Kristallogr.* **1937**, *97*, 59.
53. Thery, J.; Briancon, D. *C. R. Hebd. Seances Acad. Sci.* **1962**, *254*, 2782.
54. Schloder, R.; Mansmann, M. *Z. Anorg. Allg. Chem.* **1963**, *321*, 246-261.
55. Weber, N.; Kummer, J. T. *Advances In Energy Conversion Engineering*; ASME: New York, 1967; p 913.
56. Stevens, R.; Binner, J. G. P. *J. Mater. Sci.* **1984**, *19*, 695-715.
57. Newsome, J. W. U.S. Patent 2 642 337, June 16, 1953.
58. Torkar, K.; Krischner, H. *Monatsh. Chem.* **1960**, *91*, 757, 764-773.
59. Misra, C., Alcoa Laboratories, unpublished report, 1984.
60. Yamaguchi, G.; Yanagida, H.; Ono, S. *Bull. Chem. Soc. Jpn.* **1964**, *37*, 752-754.
61. Torkar, K.; Krischner, H. *Symp. Bauxites, Oxydes, Hydroxydes Alum., [C. R.], 1963* **1964**, *1*, 25.
62. Pentscheff, N. P.; Zaprianova, A. *Z. C. R. Hebd. Seances Acad. Sci.* **1969**, *268*, 54.
63. Saalfeld, H.; Matthies, H.; Datta, S. K. *Ber. Dtsch. Keram. Ges.* **1968**, *45*, 212.
64. *An Atlas of Alumina*; BA Chemicals Ltd., London, 1969.

3

Industrial Production of Aluminum Hydroxides

Industrial Production

Aluminum hydroxides, commonly (and incorrectly) called hydrated aluminas or alumina hydrates in the industry, are technically the most widely used and important members of the alumina chemicals family. A wide range of products is available for a large number of industrial applications. These different applications utilize physical, chemical, structural, purity, and particle size variations of the different aluminum hydroxides discussed in Chapter 2.

Production of aluminum hydroxide (gibbsite) is an intermediate stage of alumina production. Thus, the most important source of aluminum hydroxides is the bauxite refining plant. More than 94% of world alumina production is accounted for by the Bayer process for bauxite refining. However, a small amount is also produced by the "sinter process". Because of their whiteness, aluminum hydroxides produced by the sinter process are considered superior to the normal Bayer product for some applications.

The first alumina production on a commercial scale was based on the thermal decomposition of aluminum salts and sintering bauxite with soda (Le Chatelier process). One of the pioneers in this field was the Giulini plant (presently Alcoa Chemie) at Ludwigshafen, West Germany. The aluminum hydroxide produced was used mainly for conversion to alum. The demand for pure alumina increased sharply when it became the raw material for aluminum production following development of the Hall–Heroult cell in 1886. In the years following this development (1887–1888), Bayer developed the bauxite refining process that bears his name. His first work was done in an experimental plant in Yugoslavia. Subsequent travels for demonstration and installation of his process took him to Russia, several European countries including France and Germany, and the United States. In the United States Bayer worked with the Merrimac Chemical Company in Massachusetts and the Pennsylvania Salt Company. Both of these companies were suppliers of aluminum hydroxide to the chemical industry and, to a lesser degree, calcined alumina to the Pittsburgh Reduction Company, which was the predecessor of the Aluminum Company of America (Alcoa). In 1901, the Pittsburgh Reduction decided to refine Arkansas bauxite by the Bayer process to produce their own aluminum hydroxide. The company commissioned the design and construction of a Bayer refining plant at East St. Louis, IL, and the first carload of aluminum hydroxide was shipped from the East St. Louis plant on May 26, 1903.

Since this modest beginning, alumina plants have been built in more than 25 countries, and the present world capacity is over 40 million tons per year. With the development and growth of applications and markets for alumina chemicals, all the major alumina producers have, over the years, converted a part of their capacity for the production of various alumina chemicals. In fact, some of the older, smaller alumina refining plants have been totally converted to alumina chemicals production in order to remain economically viable. Presently, chemical uses account for nearly 8% of the world production.

Bauxite: The Principal Raw Material

The term bauxite is used for ores that contain economically recoverable quantities of the aluminum hydroxide minerals gibbsite, boehmite, and diaspore. The name originates from the description by P. Berthier in 1821 (1) of a sediment that occurred near the village of Les Baux in Provence, France. Early in this century commercial bauxite deposits were found in various parts of the European Alpidic mountains and also in several locations in the North American continent, such as Arkansas, Alabama, and Georgia. Since the 1920s, extensive deposits have been discovered in tropical and subtropical areas such as the Caribbean Islands, the northern areas of South America, Australia, Guinea, India, and Indonesia. The relatively small deposits in Arkansas have been depleted to poor grade after several years of commercial mining.

Aluminum hydroxide, iron oxide and hydroxide, titanium oxide, silicon oxide, and aluminosilicate minerals are the major components of all bauxites. Two principal types of bauxite ores, gibbsitic and boehmitic, are of primary interest in today's commercial processes because they contain 30–60% Al_2O_3. Gibbsite is the predominant aluminum mineral in the geologically young bauxites of the tropical climate belt. Older deposits of Europe, European Russia, Greece, and China contain mostly boehmite or diaspore. Silicon dioxide may occur as quartz. Most commonly, however, SiO_2 is associated with the clay minerals kaolinite, halloysite, or montmorillonite. These aluminosilicates react with sodium hydroxide to form insoluble sodium aluminum silicates during the extraction stage of the Bayer process; this reaction causes a loss of sodium hydroxide and alumina. The amount of the so-called reactive silica is one of the major factors that determine the quality and price of the ore. Hardness, texture, and the amount of overburden determine the methods applied for bauxite mining. Deposits in

3. INDUSTRIAL PRODUCTION OF ALUMINUM HYDROXIDES

Greece, Yugoslavia, Hungary, and Russia require deep mining to depths of several hundred meters. Tropical bauxites are frequently located so close to the surface that they can be recovered with conventional earth-moving equipment.

The proven reserves of bauxite in the world are sufficient to supply the world aluminum industry for several centuries. Total resources are estimated by the U.S. Geological Survey at 40–50 billion tons. Because of the worldwide distribution of commercially significant deposits, a disruption of bauxite supply for political reasons appears highly unlikely. In 1974, several major bauxite-producing nations formed the International Bauxite Association (IBA) with the intent of increasing control over the exploitation of their bauxite deposits. Although levies were increased substantially, competition from countries not associated with the IBA helped to maintain a reasonable price structure. Economic and political considerations currently favor location of the refining plant near the bauxite mines.

Bayer Process

The Bayer process is used almost exclusively for the extraction of alumina from bauxite. The process is considered economically suitable for bauxites containing 30–60% Al_2O_3 as aluminum hydroxides and less than 7% SiO_2 as clay (kaolin) minerals.

The basic flow sheet of the Bayer process is shown in Figure 3.1. Principal features of the process have remained unchanged since its discovery, although chemical engineering developments have considerably increased the efficiency of the process and scale of operation. The essential steps of the process are extraction of aluminum hydroxide minerals in bauxite (as $NaAlO_2$) by using a hot concentrated sodium hydroxide solution, separation of insoluble impurities (mainly oxides of iron, titanium, and silicon) by decantation and filtration, and the precipitation of crystalline aluminum hydroxide (gibbsite) from the caustic aluminate solution. This step is followed by thermal dehydration (calcination at 1000–1200°C) of the washed hydroxide to produce alumina (Al_2O_3).

Specific extraction conditions depend on the type of alumina minerals present in the particular bauxite being processed. Gibbsite is extracted at lower ($\simeq 150$ °C) temperatures and caustic concentrations ($\simeq 200$ g/L NaOH), whereas boehmite can be economically extracted only above 200 °C. Extraction of diaspore requires even more severe conditions: 300 °C and 300 g/L NaOH. Bauxite extraction (digestion) is usually carried out in continuously operated steam-heated autoclaves combined with heat recovery systems.

Primary separation of the solid impurities (bauxite residue) is effected by decantation in large-diameter thickener tanks. The decanted caustic aluminate liquor is further filtered to remove suspended solids and sent to the

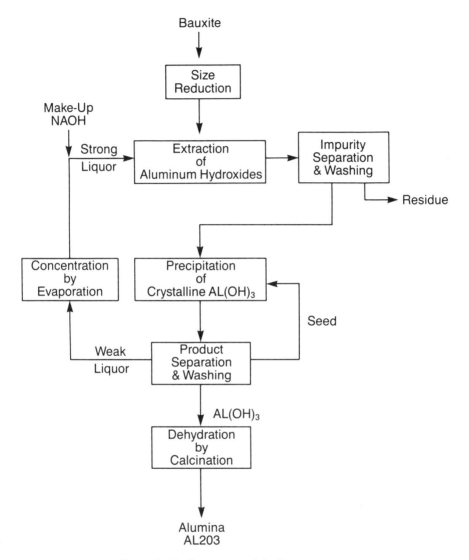

Figure 3.1A. Basic steps of the Bayer process.

hydroxide precipitation department. The dissolved alumina in the cooled (60–70 °C), supersaturated aluminate liquor crystallizes out on seed crystals of gibbsite, which are separated by filtration and washed. Crystallization mechanisms in the precipitation process involve crystal growth, agglomeration, and nucleation of the aluminum hydroxide crystals. The washed hydroxide is dried for chemical products or calcined to metallurgical- (aluminum smelting) grade alumina. The hydroxide yield in precipitation depends on the difference in solubility of alumina in the caustic liquor between ex-

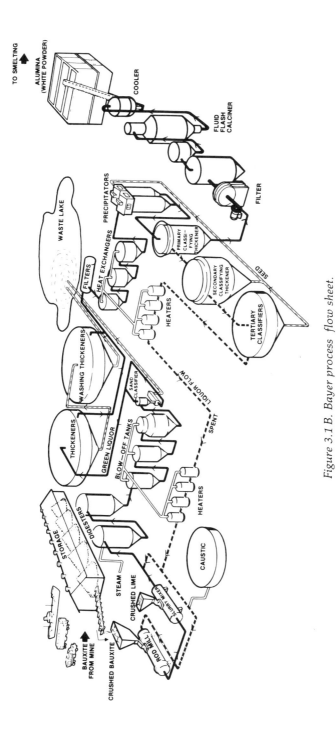

Figure 3.1B. Bayer process flow sheet.

traction and precipitation conditions. These conditions are chosen to maximize recovery and obtain a product with a desired particle size distribution. Precipitation is carried out for 30–100 h in large agitated tanks. The caustic liquor remaining after precipitation and hydrate separation is recycled back to the extraction step after concentration by evaporation. A general view of a large Bayer-process bauxite refining plant is shown in Figure 3.2. Capacities of modern Bayer plants vary from 500,000 to more than 2 million tons of alumina production per year.

During digestion, kaolin minerals in bauxite react with NaOH to form insoluble sodium aluminum silicate, which is lost with the residue. This loss represents a significant production cost factor and puts an economic limit on the amount of kaolin minerals that can be tolerated in the bauxite. Other important production cost factors are the bauxite (mining and transportation) and energy (coal, oil, and gas) used for generation of the steam and power that are used in the refining process. Table 3.I shows raw material consumption factors for alumina production by the Bayer process.

Disposal of the bauxite residue obtained from the Bayer process (0.5–2.5 tons of residue per ton of alumina) is a major environmental problem of the bauxite refining industry. This residue remains strongly alkaline even after several stages of washing and has poor consolidation characteristics. Impoundment in residue lakes has been the only economical method of disposal until now. Considerable effort is currently being expended to find improved disposal practices and to find industrial uses for the residue in the iron and steel, cement and aggregate, and construction industries. Agricultural applications, as an amendment for acidic and iron-deficient soils, continue to be investigated.

Detailed accounts of the technology of the Bayer process can be found in books by Ginsberg and Wrigge (2), Barrand and Gadeau (3), and many publications in technical journals. The general review by Adamson (4) covers principles and practices of alumina production by the Bayer process. Typical modern American Bayer practices, represented by the Point Comfort Operations of Alcoa, are described by Hoppe (5). Chemical principles of the process are discussed by Pearson (6) and Kuznetsov and Derevyankin (7).

The aluminum hydroxide obtained from the Bayer process is the crystalline trihydroxide gibbsite. This typical Bayer process hydroxide product is shown in Figure 3.3. The particle size varies from 5 to 180 μm; particles appear as roughly spherical agglomerates of hexagonal prisms and rods. The Bayer product is normally off-white, the yellow color having originated from organic matter extracted from bauxite by the caustic liquor. Chemical analyses of typical products are given in Table 3.II. The product is nearly 99.5% pure; the major impurity is soda (Na_2O). Soda is universally present in Bayer hydrate and may originate from various sources: (1) caustic liquor adhering to the surface, (2) liquor inclusions inside the agglomerates, (3) soda adsorbed on the surface and sealed in by crystal growth, and (4) soda

3. INDUSTRIAL PRODUCTION OF ALUMINUM HYDROXIDES

Figure 3.2. General view of a large Bayer-process bauxite refining plant.

Table 3.I.
Consumption Factors for Bayer-Process Alumina Production

Item	Factor
Bauxite (tons)	2–3.5
Caustic (as NaOH) (tons)	50–100
Energy (GJ)	
For hydroxide	6–10
Calcination to Al_2O_3	3.2–4.5
Quick lime (as CaO) (kg)	30–60

NOTE: Data are for the production of 1 metric ton of calcined alumina. Miscellaneous items consumed were grinding media, filter cloth, acids (for cleaning), and flocculent (for residue).

× 320

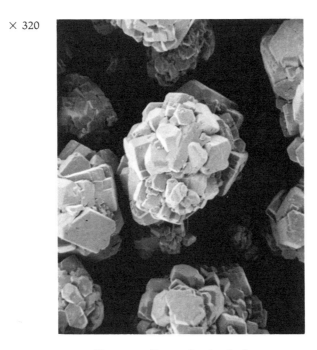

Figure 3.3. Bayer alumina hydrate.

as a constituent of the gibbsite structure. The dried hydroxide is a free-flowing powder and is transported in paper bags or in rail and road bulk carriers.

Sinter Process

Even though the soda sinter process is of relatively minor commercial significance compared to the Bayer process, it continues to be of practical interest

Table 3.II.
Typical Properties of Alcoa Series C-30 Aluminum Hydroxides

Typical Properties	C-30	C-31	C-31 Coarse	C-37
Al_2O_3 (%)	64.90 (64.5 min)	65.0 (64.5 min)	64.90 (64.5 min)	64.2
SiO_2 (%)	0.012 (0.040 max)	0.010 (0.020 max)	0.010 (0.020 max)	0.070
Fe_2O_3 (%)	0.015 (0.040 max)	0.004 (0.005 max)	0.006 (0.008 max)	0.002
Na_2O (% total)	0.400 (0.600 max)	0.150 (0.200 max)	0.200 (0.350 max)	0.420
LOI (% total)[a]	34.5	34.5	34.5	34.5
Moisture (% at 110 °C)	0.10 (0.20 max)	0.04 (0.10 max)	0.04 (0.10 max)	0.10
Bulk density, loose (g/cm³)	1.2–1.4	1.0–1.1	1.1–1.3	0.80–1.0
Bulk density, packed (g/cm³)	1.5–1.7	1.2–1.4	1.4–1.6	1.0–1.1
Specific gravity	2.42	2.42	2.42	2.53
Mohs' hardness	2.5–3.5	2.5–3.5	2.5–3.5	2.5–3.5
Surface area (m²/g)[b]	0.10	0.15	0.10	0.2
Refractive index	1.57	1.57	1.57	1.58
Color	off-white	white	white	off-white
Particle size analysis[c]				
% on 100 mesh	5–20	0–1	0–10	6–12
% on 200 mesh	65–90	5–15 (max)	40–80	30–60
% on 325 mesh	90 (min)–98	30–65	85 (min)–97	75–95
% through 325 mesh	2–10	35–70	3–15	5–25

[a]Weight loss due to heating from 110 to 1100 °C.
[b]Surface area measured by a Perkin-Elmer-Shell sorptometer.
[c]Tyler Standard Screen Series. The particle size values are cumulative.

for processing low-grade bauxites. First carried out on a commercial scale by Le Chatelier (1875) in France, the process is based on the reaction

$$Al_2O_3 + Na_2CO_3 \rightleftharpoons 2NaAlO_2 + CO_2$$

This reaction occurs at a relatively high temperature of 900–1100 °C, which converts all aluminum hydroxide minerals to $NaAlO_2$ quantitatively.

In this process bauxite is sintered with soda (Na_2CO_3) in rotary sintering kilns. Alumina minerals in the ore react with soda to form $NaAlO_2$, which is extracted by water leaching. The leach liquor then passes through residue separation and aluminum hydroxide precipitation stages similar to those of the Bayer process. The liquor is then gassed with CO_2 to precipitate out remaining aluminum hydroxide and to convert NaOH to Na_2CO_3. The liquor requires additional desilication process operations to remove silica that would otherwise contaminate the final precipitated hydroxide. Desilication is carried out by heating the liquor in autoclaves to 150–200 °C, which leads to the crystallization of dissolved silica as crystalline sodium aluminum silicate. Reaction with lime (CaO) is also employed to remove silica as calcium aluminum silicate.

The high energy consumption required for sintering makes this process uneconomical compared to the Bayer process. However, this process is capable of yielding a very white product because all the organic matter is destroyed during sintering. The whiteness of the hydroxide is a desirable feature in some industrial uses. The process is reported to be in use in Russia, Czechoslovakia, and China for the processing of poor-grade diasporic ores that normally cannot be processed by the Bayer method. A detailed description of an operating sinter plant is presented by Tomka (8). The scale of sinter operations today is much smaller compared to that of modern Bayer plants. Total world production of alumina by the sinter process is only a small percentage compared to that of Bayer production.

Combination Process

The combination Bayer–sinter process was developed commercially by Alcoa in the United States during the Second World War for processing Arkansas bauxite high in silica (kaolin) content.

In this process the high-silica bauxite is first subjected to a Bayer extraction step. The residue obtained contains appreciable amounts of soda and alumina in the form of sodium aluminum silicates. This residue is next processed in a sinter operation with the addition of lime and additional Na_2CO_3. Lime reacts with silica to form dicalcium silicate. Sodium aluminate is leached out from the sinter mass. This process thus recovers soda and alumina from the bauxite residue. The leach liquor passes through a desilication process step and is then used for aluminum hydroxide precipitation. The organic-free white hydroxide obtained from the sinter liquor is in special demand for chemical-grade products used in the paper and plastics industries.

Both Alcoa and Reynolds operate combination process refining plants located in the state of Arkansas. Details of the flow sheet and equipment are presented by Edwards (9) and Gould (10). The Reynolds plant is reported to have ceased production in 1984 for economic reasons. Production at the Alcoa plant is around 300,000 tons per year, which is mostly sold as chemical products. The combination process is reported to be currently in use in several plants in Russia.

Bayer Hydroxide Product Variants

The large size of the alumina industry allows economic production of chemical-grade alumina products as a side-stream operation. A convenient source is the intermediate aluminum hydroxide, a part of which is dried and used as chemical products. However, the operating conditions and practices of the alumina refining plant are generally geared to the cost-effective production of a single grade of metallurgical alumina. The hydroxide has rather limited chemical uses in its raw form for conversion to metallurgical alumina. Some variations in particle size, purity, and other properties are generally needed before the hydroxide can be used in the wide range of chemical applications that have been developed over the years. When crystallization conditions in the precipitation step such as liquor alumina and caustic concentrations, temperature, agitation, seed hydroxide particle size, and quantity are varied, a wide variety of aluminum hydroxide grades are produced from a part of the Bayer liquor. The liquor is diverted to a different group of precipitation vessels for this purpose. Grinding, classification, and surface treatment are used to further extend the range of particle size, dispersion behavior, and surface properties of the hydroxides. The preparation of the various grades is of considerable commercial interest and is described in the following sections.

So that hydroxides can be used as chemical products, care has to be exercised in the hydroxide-drying operation to avoid exposure to high (>150 °C) temperatures, which could cause dehydration of the hydroxide and thus affect physical and chemical properties, such as surface activity and dissolution rate in acid and alkalis. A variety of dryer types have been used to reduce the free moisture content of hydroxide filter cake from 10–15% to less than 0.5% in the dried product. These include steam tube rotary, hearth type with rotary scraper, vibratory, fluidized bed, and pneumatic conveying-type dryers.

Normal (Coarse) Grade. This corresponds to the usual bulk product from the Bayer plant produced for conversion to smelting-grade alumina. Traditionally the American Bayer product is coarse (known as "sandy") compared to the European product. Chemical compositions of the two products are nearly identical (Table 3.II). Particle size distributions of American and European products are shown in Figure 3.4. The hydroxide is approximately 99.5% pure. Principal impurities are soda (Na_2O), Fe_2O_3, SiO_2, and TiO_2. Trace organic impurities originating from bauxite and present in the liquor give it a slight off-white or yellowish color. Figure 3.3 is a scanning electron micrograph (SEM) of a typical Bayer hydroxide. This low-cost material is widely used for making alum and other aluminum chemicals.

Normal-Grade White Hydroxide. As described earlier, the sinter operation applied to bauxite (sinter process) or the residue (combination process) produces a hydroxide that is completely free from organic matter and is perfectly white. A reading of more than 95% is obtained on the GE brightness scale (relative to TiO_2 as followed in the paper industry) compared to 70%

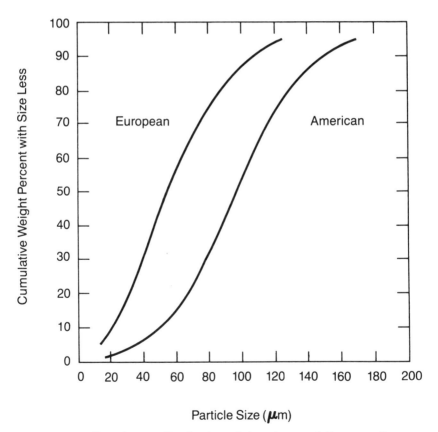

Figure 3.4. Particle size distribution of American and European Bayer hydroxide.

on the same scale for the normal Bayer product. This grade is preferred for some applications where whiteness is a desirable property. The particle size distribution and chemical analyses are otherwise nearly identical with those of the normal grade.

Ground Hydroxides. Many applications of aluminum hydroxides require smaller particle sizes. These hydroxides are produced by grinding the dry hydroxide followed, if necessary, by size classification. Various types of grinding equipment are used: ceramic-lined (to minimize iron contamination) ball and jar mills, vibrating mills, disc mills, air jet pulverizers, etc. Bayer hydroxide is relatively easy to grind, and no specific operating problems are encountered. Some soda present within the crystals is released during milling, and the resulting increase in the alkalinity of the ground product could be a problem in some applications. Particle size distributions of some typical (Alcoa C-33 series) ground hydroxides are given in Table 3.III and in Figure 3.5. Consistency of particle size distribution is a major consideration in the production of ground products.

Low-Iron Hydroxide. Some applications [e.g., production of pure (low-iron) aluminum chemicals] require a hydroxide having very low amounts of Fe_2O_3, typically 0.003% Fe_2O_3 against the normal 0.015% Fe_2O_3. This hydroxide is produced by a two-step precipitation process. Advantage is taken of the fact that the major part of iron impurity is associated with aluminum trihydroxide crystallizing out at the beginning of the precipitation cycle. The high-iron-content hydroxide is separated at this stage by decantation or filtration. The remaining liquor is then seeded with a low-iron seed hydroxide to obtain a low-iron product. Alternative procedures are to remove iron from liquor by adsorptive or chemical methods before precipitation. Limestone, lime, MgO, and calcined bauxite have been claimed to be effective for iron removal from the liquor.

Low-Soda Hydroxide. The soda (Na_2O) content of normal Bayer hydroxide is around 0.2–0.4%. Approximately 0.1% of this originates from adhering liquor, which can be removed by thorough washing. The remaining soda is trapped within the hydroxide crystal and cannot normally be washed out. Although no clear picture of the mechanisms by which this soda is occluded within the hydroxide particles is available, experience shows that soda content is reduced when precipitation is carried out at relatively high temperatures (80–95 °C) and under low-alumina supersaturation conditions. Soda contents as low as 0.05% Na_2O can be obtained by this procedure. However, precipitation at higher temperatures also reduces the recovery of hydroxide and thereby increases the cost of production. Low-soda hydroxide is generally used for the production of ceramic aluminas.

Table 3.III.
Typical Properties of Alcoa Ground Aluminum Hydroxides

Typical Properties	C-331 and C-333	C-330	C-430
Al_2O_3 (%)	65.0 (64.5 min)	65.0 (64.5 min)	65.0 (64.5 min)
SiO_2 (%)	0.010 (0.020 max)	0.020 (0.040 max)	0.020
Fe_2O_3 (%)	0.006 (0.010 max),[a] 0.004 (0.005 max)[b]	0.030 (0.040 max)	0.030
Na_2O (% total)	0.15 (0.25 max)	0.30 (0.50 max)	0.30
Na_2O (% soluble)	0.02 (0.04 max)	0.04 (0.08 max)	0.03
LOI (% total)[c]	34.5	34.5	34.5
Moisture (% at 110 °C)	0.40 (0.70 max)	0.40 (0.70 max)	0.30
Bulk density (g/cm³)	0.70–1.25	0.70–1.25	0.71–1.30
Specific gravity	2.42	2.42	2.42
Mohs' hardness	2.5–3.5	2.5–3.5	2.5–3.5
Surface area (m²/g)	5.0	5.0	4.0
Refractive index	1.57	1.57	1.57
Color	white	off-white	off-white
Particle size[d] (wt %)	98 (min)–99	98 (min)–99	87–92
Median particle size[e] (μm)	6.5–9.5	6.5–9.5	12–14

[a] C-331.
[b] C-333.
[c] Weight loss due to heating from 110 to 1100 °C.
[d] Through 325 mesh.
[e] The median particle size is measured by the Coulter counter. "Median" designates that half of the particles have sizes greater than the figure shown and half have sizes less than the figure shown.

3. INDUSTRIAL PRODUCTION OF ALUMINUM HYDROXIDES

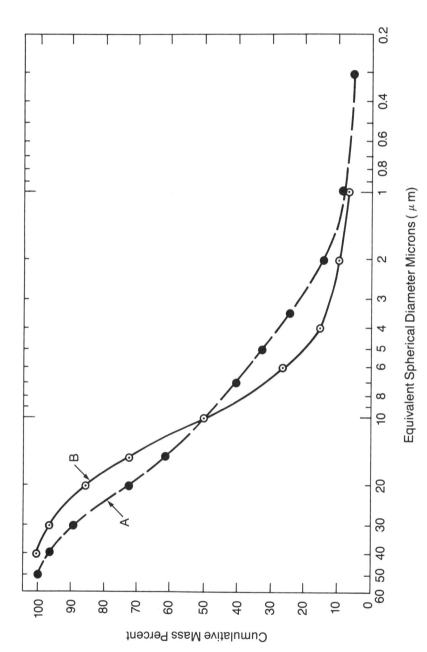

Figure 3.5. Particle size distribution of commercial ground aluminum hydroxide.

Extra-Fine Hydroxide. Extra-fine precipitated hydroxides have a uniform particle size of about 1 μm (diameter). This material is produced by precipitation under carefully controlled conditions using a specially prepared seed hydroxide. Precipitation from an organic-free aluminate liquor, such as that obtained from sinter or combination processes, yields a very white (GE brightness close to 100%) product. Precipitation is carried out at low (30–40 °C) temperatures, which causes massive nucleation of fine, uniform hydroxide particles. The product is filtered and washed thoroughly, usually on rotary vacuum filters. Conventional tray or tunnel drying of the filter cake produces a granulated product that can be easily pulverized to obtain the fine hydroxide. Oversize particles are removed by screens or air classification systems integral with the pulverizer. Alternatively, the washed product is spray-dried. The spray-dried product has a higher bulk density and improved flow characteristics compared with those of the pulverized material. The electron micrograph picture (Figure 3.6) shows the remarkable uniformity and platy characteristics of this product. Physical data relating to this product (Alcoa Hydral) are given in Table 3.IV.

Surface-Treated Hydroxides. The various hydroxide grades can be surface-treated to modify dispersion behavior and rheological properties to suit many industrial applications. The most widely used surface-coating agents are stearic acid and its salts, although other coating agents, for example, oleates and silicones, have also been used in specific applications. Coating is normally carried out batchwise in a high-shear powder-mixing machine. The rise in temperature resulting from dissipation of mixing power is often sufficient to give good dispersion, but external heat may be applied through steam jacketing or coils.

Production of Other Aluminum Hydroxides

Bayerite. The major technical use of bayerite is as a catalyst base material. Bayerite is produced commercially by CO_2 neutralization of caustic aluminate liquor obtained from either Bayer or sinter processes. Neutralization is carried out by gassing with clean flue gas containing 10–15 vol % CO_2 under controlled conditions of gas flow rate, temperature, and agitation. Pure CO_2 has also been used. The product obtained is about 90% crystalline bayerite with minor amounts of gibbsite, pseudo-boehmite, and amorphous aluminum hydroxides. The product is filtered, washed, and dried. Typical characteristics of the product (Alcoa C-37 hydroxide) are shown in Table 3.II. The SEM photograph (Figure 3.7) shows the different morphology of this product compared to that of the usual Bayer gibbsite.

Boehmite. Three different boehmite products are of commercial significance.

3. INDUSTRIAL PRODUCTION OF ALUMINUM HYDROXIDES

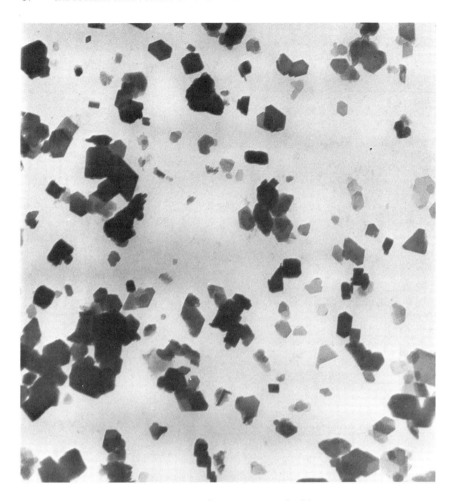

Figure 3.6. Extra-fine precipitated gibbsite.

HYDROTHERMALLY PRODUCED BOEHMITE. This product is obtained by hydrothermal conversion of Bayer hydroxide in pressure autoclaves in the presence of water. The process reduces the alkali content of the final product, but other impurities present in the starting hydrate are retained. Various particle size products are obtained by starting from different particle size hydroxides and also by grinding the product. The product comprises agglomerates of small, approximately 5-μm size, well-crystallized boehmite crystallites in the form of platelets (Figure 3.8); the size of these crystallites is controlled by the temperature and pressure used for hydrothermal conversion. Crystallite size is also affected by the concentration of alkali during hydrothermal conversion; higher alkali concentrations produce finer crystallites. The platelets can be released by intensive grinding. One manufac-

Table 3.IV.
Typical Properties of Alcoa Hydral Series Aluminum Hydroxide

Typical Properties	Hydral 710	Paper Grade (Spray-Dried)
Al_2O_3 (%)	64.7	64.7
SiO_2 (%)	0.04	0.04
Fe_2O_3 (%)	0.01	0.01
Na_2O (% total)	0.45	0.45
Na_2O (% soluble)	0.10–0.25 (max)	0.10–0.25 (max)
Moisture (% at 110 °C)	0.3–1.0 (max)	0.3–1.0 (max)
Bulk density, loose (g/cm^3)	0.13–0.22	0.35
Bulk density, packed (g/cm^3)	0.26–0.45	0.70
Specific gravity	2.42	2.42
Surface area (m^2/g)[a]	6–8	6–8
Mohs' hardness	2.5–3.5	2.5–3.5
Refractive index	1.57	1.57
Color	white	white
GE brightness (%)	94+	94+
Particle size analysis[b]		
% on 325 mesh	0.15 (max)	0.15 (max)
% less than 2 µm	100	—
% less than 1 µm	85	—
% less than 0–0.5 µm	28	—

[a]Surface area measured by the Brunauer–Emmett–Teller method of nitrogen adsorption.
[b]As determined by using an electron microscope on a weight basis. The particle size values are cumulative.

turer of this product (BACO "Cera" hydroxide) recommends applications in ceramic and polishing formulations.

BYPRODUCT OF LINEAR ALCOHOL PRODUCTION. The Ziegler process for synthesizing linear alcohols involves the formation of aluminum alkoxides as an intermediate product. Hydrolysis of aluminum alkoxides produces linear alcohols and the fine aluminum hydroxide having the crystal structure of boehmite. The X-ray diffraction pattern of this material shows broad bands similar to that of pseudo-boehmite. The hydrate slurry is further processed to remove residual alcohols and then dried in either rotary or spray dryers. Typical physical and chemical properties of this product are given in Table 3.V. The product is marketed under the trade name of Catapal and has been claimed to be an excellent catalyst base material, in addition to other applications.

FIBROUS BOEHMITE. Fine-particle fibrous boehmite was available for some time as a commercial product, Baymal, produced by Du Pont. Fibrous boehmite is prepared by hydrothermal conversion of aluminum salts (e.g., alumi-

3. INDUSTRIAL PRODUCTION OF ALUMINUM HYDROXIDES   49

× 70

× 700

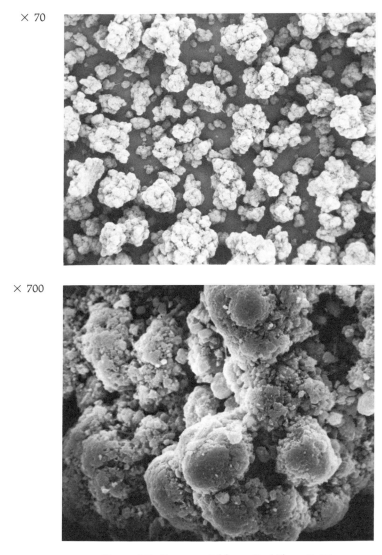

Figure 3.7. Commercial bayerite (Alcoa C-37).

num sulfate) in the presence of acetic acid. The product, after filtration and drying, is a free-flowing white powder consisting of clusters of minute fibrils of boehmite. The powder disperses readily in water to yield sols of ultimate fibrils. The surface of the fibrils is modified by adsorbed acetate ions, which play an important role in the colloidal behavior and many useful properties of the product. Typical chemical compositions and properties are given in Table 3.VI.

× 160

× 1600

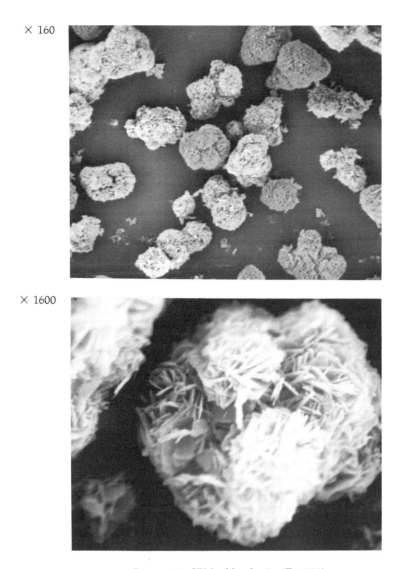

Figure 3.8. SEM of boehmite (D-6883).

Gelatinous Aluminum Hydroxides. Commercial production of gelatinous aluminum hydroxides generally follows two different routes:

> 1. A gel is obtained by neutralization of aluminum salts (sulfate, nitrate, chloride, etc.) with ammonium or sodium hydroxides. The resultant product is often aged to improve washability and washed thoroughly to remove anions. However, the gelatinous product is difficult to filter and wash, and normally contains large amounts of anions. The

3. INDUSTRIAL PRODUCTION OF ALUMINUM HYDROXIDES   51

Table 3.V.
Typical Properties of Boehmite from Linear Alcohol Production

Property	Catapal SB	Catapal NG	Dispal M
Crystal structure[a]	boehmite	boehmite	boehmite
Crystal structure[b]	γ-alumina	γ-alumina	γ-alumina
Surface area[b] (m^2/g)	250	180	185
Crystallite size (X-ray) (Å)			
020 reflection[a]	34	65	60
021 reflection[a]	48	100	95
440 reflection[b]	45	49	49
Pore volume (cm^3/g)			
0–100 Å [b]	0.45	0.40	0.40
0–10,000 Å [b]	0.50	0.45	0.45
Helium density (g/cm^3)[b]	3.32	3.32	3.32
Loose bulk density[a] (g/L)	790	920	610
Particle size distribution[a]			
% 45 μm	48	73	86
% 90 μm	9		
% 200 μm			2
Dispersibility[c] (%)			98.5
Typical chemical composition (wt %)			
Al_2O_3 content	74.2	75.8	73.1
Total ignition loss	25.8	24.2	26.9
Carbon[d]	0.36	0.30	0.30
Silica (as SiO_2)	0.008	0.008	0.008
Iron (as Fe_2O_3)	0.005	0.005	0.005
Sodium (as Na_2O)	0.004	0.004	0.004
Sulfur	0.01	0.005	0.005

[a]Plant product.
[b]After calcination for 3 h at 400 °C.
[c]4% HNO_3.
[d]As primary alcohol; removed during calcination.

gelatinous mass is converted to hard lumps on drying, which are pulverized to a fine powder. Alternatively, the gelatinous material can be extruded to form pellets and then dried. Spherical particles of dried gel have been produced by controlled addition of the gel to hot oil.

2. A sodium aluminate solution is neutralized by acids, $NaHCO_3$, or CO_2 to yield a gelatinous hydroxide. The product is filtered, washed, and dried. Again, the filtration and washing operations are difficult; often a high sodium content remains in the product.

Gelatinous aluminum hydroxides are extensively used in the preparation of adsorbent and catalytic aluminas. They are also used in the pharmaceutical industry. Wolfe (11) has given an account of the production of gelatinous aluminum hydroxide used in the manufacture of printing inks. The patent literature contains a large number of examples describing the preparation of alumina gels of varying properties and applications. Most of these processes

Table 3.VI.
Baymal Fibrous Boehmite

Typical Chemical Composition	
Component	wt %
Major	
AlOOH	88.10
CH$_3$COOH	9.80
SO$_4$	1.70
Water	5.00
Minor	
NH$_4$	0.02
Na	0.07
Fe	0.02
SiO$_2$	0.02

Typical Physical Properties	
Property	Value
Specific surface area (m^2/g)	274
Pore volume (cm^3/g)	0.53
Pore diameter (Å)	77
Bulk density (lb/ft^3)	
Loose	26
Packed	31
Absolute density (fibril) (g/cm^3)	2.28
Refractive index (fibril)	1.580[a]
Oil absorption	147[b]
Color	white
pH (4% sol)	
With KCl bridge–calomel cell	3.8
No bridge	4.3
Particle charge in sol	positive

[a] n_{25}.
[b] ASTM method D-281-31. Standard test method for oil absorption of pigments by spatula rub-out. ASTM, Philadelphia, 1981.

involve neutralization of aluminum salts or aluminates. Product variations are often related to the particle size of the colloidal hydroxide and the forming method used. Rate, temperature, pH, and agitation during neutralization and subsequent aging to different degrees of crystallinity all play significant roles in the development of reproducible properties of the final product.

Literature Cited

1. Berthier, P. *Ann. Mines* **1821**, *6*, 531–534.
2. Ginsberg, H.; Wrigge, Fr. W. *Tonerde und Aluminium, Teil-1, Die Tonerde*; de Gruyter: Berlin, 1964.
3. Barrand, P.; Gadeau, M. *L' Aluminium (Part 1)*; Editions Eyrolles: Paris, 1964.
4. Adamson, A. N. *Chem. Eng. (London)* **1970**, *1*, 156–171.

5. Hoppe, R. *E/MJ Second Operating Handbook of Mineral Procedures;* McGraw-Hill: New York, 1979; pp 133–137.
6. Pearson, T. G. *The Chemical Background of the Aluminum Industry;* Lectures, Monographs and Reports No. 3; The Royal Institute of Chemistry: London, 1955.
7. Kuznetsov, S. I.; Derevyankin, V. A. *The Physical Chemistry of Alumina Production by the Bayer Process;* Metallurgical Press: Moscow, 1964.
8. Tomka, L. *Proc. ICSOBA Conf. Czech.* **1972**, 195–212.
9. Edwards, J. D. *AIME Tech. Publ. No. 1833* **1945**.
10. Gould, R. F. *Ind. Eng. Chem.* **1945**, *37*, 796–802.
11. Wolfe, H. J. *Printing and Litho Inks;* MacNair-Dorland: New York, 1968.

4

Aluminum Hydroxides: Industrial Applications

Aluminum hydroxides constitute a group of versatile industrial chemicals. Their uses are varied and cover a wide range of industries. Important uses requiring large tonnages are as fillers in plastics and polymer products and for the production of aluminum chemicals. A modest amount is used for the production of alumina-based adsorbents and catalysts. In the following sections, I will discuss these and other industrial applications of aluminum hydroxides.

Filler in Plastics and Polymer Systems

Aluminum hydroxide from the Bayer process is an inorganic mineral product that is finding increasing application as a filler in plastics and polymer systems. Aluminum hydroxide has all the requirements to be an effective filler: white or near-white color, large-volume production base ensuring dependable supply, consistency of physical and chemical properties, availability in a wide range of particle size distributions, chemical inertness, nontoxicity, and, finally, cost-effectiveness. Its use as a filler is, however, strongly related to its fire-retardant and smoke-suppressant properties that tend to compensate for its somewhat higher price compared to that of calcium carbonate and other mineral fillers.

The combustion of a polymeric material exposed to an external source of heat occurs in several stages: (1) heating, (2) decomposition, (3) ignition, (4) combustion, and (5) propagation.

In the heating stage, the rate of temperature rise is dependent upon the thermal characteristics of the polymer: specific heat, thermal conductivity, and latent heats of phase changes. Once the material reaches a high enough temperature, decomposition begins, usually accompanied by the evolution of several products: combustible gases such as carbon monoxide, hydrogen, and low molecular weight organic compounds; noncombustible gases such as water vapor and hydrogen chloride; and solids in the form of char and smoke particles. Combustion begins when the combustible gaseous products

0065-7719/86/0184/0055$06.00/1
© 1986 American Chemical Society

ignite, either as a result of a spark or nearby flame or, spontaneously, at a somewhat higher temperature. During combustion, the heat evolved by the burning gaseous products raises the temperature of the surrounding material and may also bring them into combustion; thus, the fire is propagated.

All fire retardants work by interfering with one or more stages of this combustion process. Aluminum hydroxide operates primarily in the first two stages of the combustion process: heating and decomposition (1). Aluminum hydroxide decomposes above 200 °C to give alumina and release water vapor by the reaction

$$2Al(OH)_3 \longrightarrow Al_2O_3 + 3H_2O$$

This reaction is strongly endothermic, absorbing approximately 280 cal/g of hydroxide. The maximum decomposition rate occurs between 300 and 350 °C, which roughly corresponds to the range of decomposition temperature of most commonly used polymers. Thus, when a polymer containing aluminum hydroxide is heated, the aluminum hydroxide decomposes endothermically, acts as a heat sink, and slows the rate of temperature rise and the rate of decomposition of the polymer. As a secondary effect, the water vapor given off also serves to dilute any combustible gases evolved; thus, their ignition is more difficult. The thermal behavior of aluminum hydroxides is discussed in detail in a subsequent chapter.

The smoke-inhibiting activity of aluminum hydroxide fillers has been attributed to its promotion of solid-phase charring instead of soot formation. Smoke suppression is also probably related to endothermic dehydration, because heat dissipation in the condensed phase reduces pyrolysis and favors competing cross-linking reactions.

With a Mohs hardness of 2.5, aluminum hydroxide (gibbsite) is only moderately abrasive; thus, wear of the processing machinery is minimized. Because gibbsite is a crystalline product, its surface has low sorptive capacity and hence does not increase resin demand in liquid polyester, epoxy, and acrylic systems. When used as a filler in plastics exposed to electric arcing, gibbsite contributed to increased arc and track resistance.

Although offering these desirable features, aluminum hydroxide has some disadvantages that impose some limitations on its use as a filler. Like many other nonreinforcing mineral fillers, aluminum hydroxide generally lowers strength properties. Because it undergoes thermal decomposition, aluminum hydroxide is not suitable for processing above the temperature of about 200 °C. The decomposition behavior, shown in Figure 4.1, shows rapid water release above 220 °C. The soluble soda content of Bayer hydroxide has been shown to cause slow degradation, property alteration, and processing problems in some polymer systems.

Although Bayer aluminum hydroxide is slightly more expensive than other mineral fillers, its cost advantage as a fire-retardant filler has contrib-

4. INDUSTRIAL APPLICATIONS OF ALUMINUM HYDROXIDES

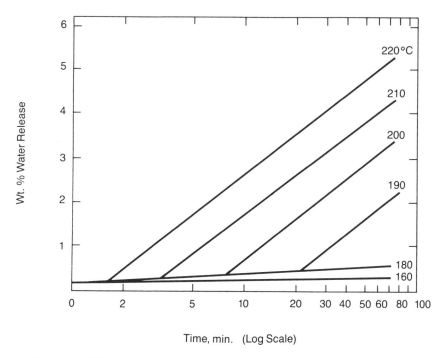

Figure 4.1. *Effect of temperature and time on water release from 20-μm aluminum trihydroxide (Bayer product).*

uted to its popularity. The material is relatively low in cost compared to other flame retardants such as antimony and bromine compounds. Aluminum hydroxide is easy to handle, is nontoxic, is free of halogens, and does not produce corrosive or toxic decomposition products. Unlike many organic additives, aluminum hydroxide is not volatile, is odorless, and is chemically inert. White grades that do not discolor on aging are available. The low decomposition temperature (approximately 200 °C) of aluminum hydroxide has, however, been responsible for its greater use as a filler in latex carpet backings and in glass-reinforced polyesters where processing temperatures below about 205 °C are generally used. These usages are also chemically cross-linked systems or products with fiber reinforcements, and the loss of strength caused by the use of a nonreinforcing filler may be acceptable for such materials.

Significant strength reduction occurs in thermoplastics as hydroxide loading is increased, and this reduction combined with the temperature constraint has been largely responsible for restricted usage in this area of applications. Efforts have been made to improve strength properties of filled plastics by the use of coupling agents such as silanes. The silane bonds the inorganic filler to the organic polymer; increased physical strength results.

The various polymer systems where aluminum hydroxide has been successfully employed as a filler are given in the box below. Although the flame-retardant properties of aluminum hydroxide were reported earlier (2), the material became a popular filler by the late 1960s. Its use gained considerable momentum with the passing of consumer safety-oriented legislation in the United States and other countries that made it mandatory for products such as synthetic rubber carpet backings, automobile furnishings, etc., to possess a certain degree of flame retardancy. Aluminum hydroxide then began to replace calcium carbonate as a filler in these products. The next major market for aluminum hydroxide was in glass-reinforced unsaturated polyesters, and here again developments closely followed the passage of flammability regulations. Public awareness and demand for safety in plastic usage should continue to support the growing use of aluminum hydroxide fillers in the future. The estimated U.S. consumption of aluminum hydroxide as a filler in rubber and plastics was around 350,000 tons in 1982. The major market for this material continues to be the carpet industry, where it is used as a filler in latex foams and adhesives for carpet backing. Approximately 200,000 tons was used for this application in 1982. About 140,000 tons was used in plastics, a large part of which went into glass-fiber-reinforced polyester products. Vinyls, epoxies, and polyolefines consumed smaller amounts. Polyurethanes and poly(vinyl chloride) (PVC) are considered to be potentially major application areas in the future.

Following is a brief description of some important filler applications.

Latex Foams and Adhesives for Carpet Backing. Carpet backing was one of the earliest large-volume applications of aluminum hydroxide as a flame-retardant filler and still remains a significant consumer.

Actually, two latex applications occur in the carpet industry, and aluminum hydroxide is used in both to impart fire retardancy. A precoat of carboxylated latex acts as an adhesive. Carpet fibers (tufts) are punched through a web backing and are anchored by the carboxylated latex. The

---

Polymer Systems Using Aluminum Hydroxide as a Filler

Thermoset	*Thermoplastic*	*Elastomeric*
Epoxy	Acrylic	Neoprene
Phenolic	PVC	SBR
Polyester	Polyethylene	Natural rubber
Spray up	Polypropylene	Butyl rubber
Hand lay up		Nitrile rubber
Foam		EPDM
SMC		Silicone rubber
BMC		
Polyurethane		

latex layer is then sandwiched between another piece of woven material (jute or polypropylene) to complete the primary backing. A secondary backing, either a separate pad or attached foam, finishes the carpeting. The secondary backing lends cushioning and insulation to the carpeting. The latex foam is transferred from a compounding and foaming unit to the precoated carpet and applied to the desired thickness. Cure is completed in an oven at 120–150 °C.

The foam compounds normally contain 20–150 parts of hydroxide/100 parts of latex, while the adhesive compounds contain 75–250 parts (*3a*). The hydroxide is usually a ground product with particle sizes in the range of 15–45 μm. Use of aluminum hydroxide as the filler in carpet backing, in place of other mineral fillers such as calcium carbonate or clay, allows the latex-backed carpet to pass the "pill test" (DOC FF 1-70) (*3b*). Basically, the pill test is designed to measure ease of ignition from a small source such as a burning cigarette. In the test a pill of methenamine is ignited under standardized conditions and the carpet area charred by the flame is measured.

Reports of "breaking" or collapse of the latex foam during processing caused by the aluminum hydroxide have occurred. A patent awarded to Geppert and Woosley (*4*) gives a method of steam treatment of the ground hydroxide immediately after grinding that avoids this problem.

Unsaturated Polyesters. The very large number of publications relating to the use of aluminum hydroxide filler in polyester systems is indicative of the volume and importance of this application (*5–11*).

Comparison of the thermal behavior using differential scanning calorimetry (DSC) of a general-purpose polyester resin with and without aluminum hydroxide is shown in Figure 4.2. The polyester resin shows little change in heat content until a temperature of about 220–230 °C is reached when an exothermic oxidative degradation process begins with the formation of low molecular weight partially oxidized combustible products. The same resin filled with 60% aluminum hydroxide shows the onset of oxidative degradation at 230 °C. But at slightly higher temperatures (250–300 °C), decomposition of the hydroxide becomes significant. Absorption of heat by the hydroxide significantly decreases heat flux to the resin. The smoke-suppressive role of aluminum hydroxide is shown by the relative smoke ratings given in Table 4.I. Aluminum hydroxide-filled resin produced far less smoke than the other samples.

In order to impart significant flame-retardant properties to polyesters, relatively high loadings of aluminum hydroxide must be used. With incremental additions of aluminum hydroxide, fire retardance increases slowly until about 40–50% hydroxide loading and then much more rapidly with further additions (Figure 4.3). The viscosity of the starting mixture increases with increased loading, and the physical properties of the cured composites also change. As with inert mineral fillers, stiffness generally increases with in-

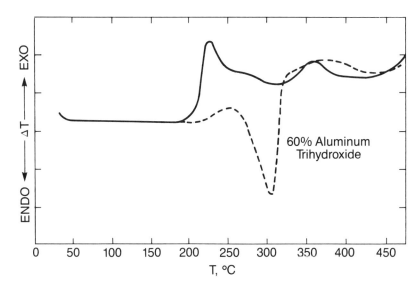

Figure 4.2. DSC of GP polyester resin alone and filled with 60% aluminum hydroxide. Reproduced with permission from Reference 10. Copyright 1974, Alcoa.

Table 4.I.
Comparison of Smoke Generated from Flame-Exposed Polyester Resin Systems

Resin	Additive	Additive (%)	Oxygen Index (% O_2)	Relative Smoke[a]
GP	Al(OH)$_3$	60	38	1
GP	CaCO$_3$	60	20	7
Chlorinated	Al(OH)$_3$	40	38	4.5
Chlorinated	Sb$_2$O$_3$	5	38	12

[a]Calculated from the weight of resin burned and the amount of smoke produced.

creased hydroxide loading while impact and tensile strength decrease. Total formulations (resin, filler, and glass fibers) need to be optimized for the desired balance between fire retardance, physical, and strength properties.

Laminate strengths are affected very little by the particle size of the hydroxide (3). Flame retardance is also relatively independent of hydroxide particle size, except for very fine grades (i.e., less than about 4-μm average diameter), which function slightly more efficiently at equivalent loadings (3, 10). Use of such fine grades in the necessary concentrations is, however, often precluded by the resulting high viscosity of the mixture.

For spray-up applications, average particle size in the range of 6–14 μm proved to be optimum. Coarser grades have a tendency to settle in process tanks and piping. Viscosity depressants such as triethyl phosphate or di-

4. INDUSTRIAL APPLICATIONS OF ALUMINUM HYDROXIDES

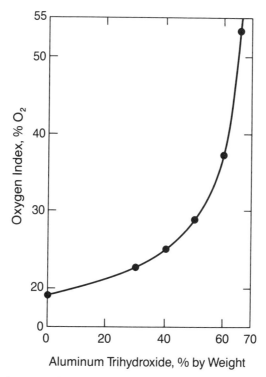

Figure 4.3. Oxygen index versus Al(OH)$_3$ concentration in polyester resin.

methyl ethyl phosphate are sometimes added to permit higher loadings without exceeding manageable viscosities.

Epoxies. Martin and Price (12a) have reported the effectiveness of aluminum hydroxide as a flame-retardant filler in epoxy resins. Used at loadings of 40–60% of resin weight, aluminum hydroxide results in a sharp increase in the oxygen index. [Oxygen index is the minimum volume percent oxygen in a slowly rising mixture of oxygen and nitrogen that is just able to support combustion of a material burning in a candlelike manner (ASTM D2863-70) (12b).] In electrical applications of epoxy resins, improvement of arc and arc-track resistance brought about by aluminum hydroxide is perhaps more important than flame retardancy. Cycloaliphatic epoxy resins filled with aluminum hydroxide have demonstrated superior electrical performance in such applications as insulators, transformers, and switch gear (13, 14). Best performances in improving electrical properties result through the use of a fine aluminum hydroxide with a low soda content. Bruins (15) has pointed out that the soluble soda (soluble in the epoxy resin) is apparently more significant than total soda in affecting electrical properties of the filled product.

Acrylics. A relatively recent, but rapidly growing application of aluminum hydroxide filler is in cast cross-linked acrylics. Apart from its flame-retardant function, the aluminum hydroxide filler imparts a marblelike translucency to the cast acrylic, and these products are marketed as "synthetic marble" for various consumer applications such as kitchen counters, bathroom panels, etc. Translucency is related to the relative difference in the refractive indices of the aluminum hydroxide (gibbsite) and the polymer. A white hydroxide (from sinter processes) is preferred in this application to accentuate the marblelike appearance.

Other Thermosets. The patent literature contains several examples of the use of aluminum hydroxide in polyurethanes, particularly foams (16-19). Bonsignore (20) has advocated the post-treatment approach in the application of aluminum hydroxide in achieving flame retardancy. In this approach, aluminum hydroxide is combined with a polymeric binder, which is applied as an aqueous latex dispersion to the foam followed by squeeze-out, drying, and curing. Problems and prospects for application in rigid high-density polyurethane foams have been analyzed by Bonsignore and Levendusky (21).

Examples of the use of aluminum hydroxide in polyethylene (22, 23), ethylene–vinyl acetate copolymers (24), diallyl phthalate molding compounds (25), and melamine formaldehyde molding compounds (26) have also been reported.

Thermoplastics. As indicated earlier, the decomposition of aluminum hydroxide starting around 200 °C has restricted its general use in thermoplastics. However, its application in thermoplastics is attractive in those cases where nonreinforcing fillers can be used and where processing temperatures are lower than 200 °C. A suggestion has been made for the use of boehmite for higher temperature applications. Boehmite decomposes at a temperature of above 500 °C and has some fire-retardant effect, although not so marked as for the trihydroxide.

Among thermoplastics, plasticized PVC has been the chief candidate for incorporation of an aluminum hydroxide filler (27). A high level of flame retardancy and smoke suppression is achieved by replacing the mineral (e.g., $CaCO_3$) filler with aluminum hydroxide and using a phosphate plasticizer. The effect of increased hydroxide content on the oxygen index of a phthalate-plasticized PVC is shown in Figure 4.4.

In rigid PVC, which has inherently very low flammability, use of aluminum hydroxide as a filler has been investigated not so much for any fire-retardant effect but rather for its ability to reduce smoke and absorb the HCl gas evolved during burning. The effect of aluminum hydroxide on the flammability, smoke generation, and strength of other thermoplastics (polyeth-

4. INDUSTRIAL APPLICATIONS OF ALUMINUM HYDROXIDES

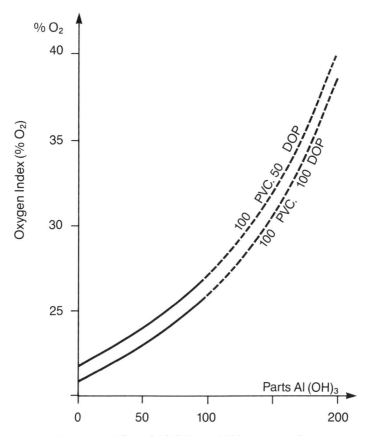

Figure 4.4. Effect of $Al(OH)_3$ on PVC oxygen index.

ylene, polypropylene, and polystyrene) is summarized in Table 4.II. As can be seen, both flame-retardancy and smoke-suppression effects have been obtained with aluminum hydroxide. However, as expected, the use of fillers has resulted in lowering the tensile strength of these materials.

Paper Applications

Fine, precipitated aluminum hydroxide having a uniform particle size of about 1 μm is used in paper making both as a filler pigment and for coating applications.

As a filler aluminum hydroxide disperses rapidly with low sedimentation. Retention in the stock is normally greater than 95% on account of the strong affinity of the hydroxide for cellulose. Its range of application is wide, from light (9 lb of tissue) to heavy (80 lb of offset and litho) papers. The

Table 4.II.
Effect of Aluminum Hydroxide on the Flammability and Strength of Thermoplastics

Filled Polymer[a]	Oxygen Index (% O_2)	UL 94 Horiz. Burn Rate (in./min)	Smoke Density Rating (%)	Tensile Strength (lb/in^2)	Impact[b] Strength (ft-lb/in)
Polyethylene, unfilled	—	0.79	4	3250	0.41
Polyethylene + CaCO$_3$	—	0.95	4	2940	0.31
Polyethylene + H-46[c]	—	SE[d]	2	2440	0.35
Polypropylene, unfilled	17.4	0.90	15	5280	0.44
Polypropylene + CaCO$_3$	19.0	0.93	5	3210	0.39
Polypropylene + H-46	22.1	0.52	6	3310	0.37
ABS[e], unfilled	—	1.64	89	5760	4.30
ABS + CaCO$_3$	—	1.31	82	3300	0.52
ABS + H-46	—	0.78	57	3280	0.50
Rigid PVC, unfilled	—	SE	89	7420	0.46
Rigid PVC + CaCO$_3$	—	SE	71	4340	0.60
Rigid PVC + H-46	—	SE	50	4180	0.50
Polystyrene, unfilled	17.9	1.43	84	3820	0.21
Polystyrene + CaCO$_3$	19.6	1.19	80	2840	0.19
Polystyrene + H-46	22.6	1.23	78	2910	0.21

SOURCE: Reproduced with permission from Reference 38. Copyright 1978, Van Nostrand-Reinhold. Data are from Sobolev and Woycheshin (28), which should be consulted for further details on materials and methods.
[a]The filler content was 40%. The aluminum hydroxide was Kaiser Aluminum's H-46 with a median particle diameter of 20 µm; the calcium carbonate was Atomite, from Thompson, Weinman & Company.
[b]Notched Izod impact.
[c]Grade of aluminum hydroxide.
[d]Self-extinguishing.
[e]Acrylonitrile–butadiene–styrene polymers.

brightness of most pulps is improved by using white-grade hydroxides that have an optical brightness of +100% (GE). In contrast to TiO_2, aluminum hydroxide does not affect the functioning of optical whiteners. However, on account of its lower refractive index, opacity attained with aluminum hydroxide is lower than with TiO_2 but is higher than with China clay. Cost advantages can be secured in many cases by partial replacement of TiO_2 with aluminum hydroxide. Improved printing properties of the hydroxide-filled paper such as high absorbance for printing inks and a reduced tendency toward dusting are also reported (29–31).

The application of fine, platy aluminum hydroxide as a paper coating was developed nearly 40 years ago and is quite well documented (32, 33). Aluminum hydroxide gives a coating of high brightness, opacity, and gloss and at the same time retains high ink receptivity. Aluminum hydroxide is normally used as full or partial replacement of China clay to improve these properties. The high-gloss effect is considered to result from the platy nature of the crystals. These crystal platelets become oriented parallel to the paper surface; this orientation results in a high degree of smoothness and gloss,

This effect is particularly evident in supercalendering. The degree of smoothness attained at relatively low pressures is similar to that achievable with other pigments only at much higher pressures, which often adversely alter other properties of the paper. The use of the hydroxide is also reported to reduce binder demand (*34*).

Raw Material for Production of Other Aluminum Compounds

Bayer aluminum hydroxide is a highly refined, 99.5% pure product. This compound dissolves readily in strong acids and alkalis. For these reasons aluminum hydroxide is the preferred base raw material for the production of a large number of aluminum compounds of both commercial and laboratory significance. The following is a summary of some of the important aluminum chemicals produced from aluminum hydroxides.

Aluminum Sulfate and Alums. Aluminum sulfate and alums are large-tonnage industrial chemicals. The paper industry and water-purification processes are major consumers of these chemicals.

Although some commercial grades of aluminum sulfate are produced directly from bauxite by dissolving the aluminum hydroxides in bauxite in strong sulfuric acid, purer, iron-free grades are normally produced from aluminum hydroxide obtained from the Bayer process. The paper industry is a large consumer of this purer, white-grade, aluminum sulfate.

Aluminum sulfate [$Al_2(SO_4)_3 \cdot 18H_2O$] is produced by reacting aluminum hydroxide with 50–70% H_2SO_4 in large acid-resistant digester tanks for several hours at temperatures of 110–130 °C. The resulting solution is concentrated by evaporation, which results in the crystallization of hydrated aluminum sulfate. The final dried product rarely contains the stoichiometric 18 H_2O but is usually slightly dehydrated (13–15 H_2O) and often contains excess alumina over the stoichiometric composition.

The term "alum" is generally applied to double salts of aluminum sulfate and another alkali metal sulfate, such as Na_2SO_4, K_2SO_4, or $(NH_4)_2SO_4$. The production process for alums is similar to that of aluminum sulfate. The second component of the alum [Na_2SO_4, K_2SO_4, or $(NH_4)_2SO_4$] is added in stoichiometric proportion to the hot aluminum sulfate solution prior to evaporation and crystallization. As with aluminum sulfate, the paper industry and water-purification processes are large users of alums.

Aluminum Fluoride. Aluminum fluoride is produced by reacting aluminum hydroxide with either hydrofluoric acid or fluorosilicic acid. The latter is a byproduct of the phosphate fertilizer industry.

In the wet process, reaction is carried out in aqueous medium. The precipitated aluminum fluoride is separated by filtration, washed, and dried at high temperatures (400–500 °C) to remove water of hydration.

Corrosion problems are claimed to be less severe in the dry process. In this process the reaction between dry aluminum hydroxide (with or without partial dehydration) and HF gas is carried out at a temperature of above 500 °C.

$$3HF + Al(OH)_3 \longrightarrow AlF_3 + 3H_2O$$

The dry process is carried out in a number of reaction stages, generally fluidized beds, to attain high conversion efficiency. A recently developed process (Lurgi process) using a circulating fluid bed is claimed to offer several advantages.

One of the principal uses of aluminum fluoride is as a component of the molten electrolyte used for the smelting of aluminum metal.

Synthetic Zeolites (Molecular Sieves). Synthetic zeolites are an important group of technical products used in gas drying and separation processes, as catalysts in the petroleum industry, and more recently as a phosphate substitute in laundry washing powders.

Synthetic molecular sieve zeolites are a class of crystalline aluminosilicate compounds with well-defined crystal structure. The general formula of this class of compounds is represented by $Me_{2/n}O \cdot Al_2O_3 \cdot XSiO_2 \cdot YH_2O$, where Me is an alkali (mostly Na or K) or alkaline earth (mostly Ca) metal and n its valency.

The crystal lattices of these compounds are such that upon dehydration they form uniform pore openings of molecular dimensions of 3–13 Å. This structure permits these compounds to accept or reject molecules depending on their size and structure, hence the name molecular sieves.

Zeolite molecular sieves containing Na are produced by controlled reaction between sodium aluminate ($NaAlO_2$) and sodium silicate in water solution. The precipitated crystalline sodium aluminum silicate is separated by filtration, washed, and dried. Activation is carried out at 400 °C to remove water of hydration. Sodium has been substituted by other metal ions by a process of ion exchange to produce other molecular sieves.

Sodium Aluminate. Sodium aluminate ($NaAlO_2$) is produced by reacting aluminum hydroxide with strong NaOH solution. The principal use of sodium aluminate is in water purification processes. This chemical is discussed in greater detail in Chapter 9.

Alumina Adsorbents and Catalysts. Various forms of aluminum hydroxides are the base material for the preparation of "activated aluminas", which are widely used as adsorbents and catalysts. This subject is discussed in considerable detail in later chapters of this book.

Abrasives, Ceramics, and Refractories. Aluminas obtained by calcining aluminum hydroxide at temperatures between 500 and 1800 °C are used in large amounts in the abrasive, ceramic, and refractory industries. A comprehensive review of these applications is available in the book *Alumina as a Ceramic Material* by Gitzen (35). A condensed version is given by MacZura et al. (36).

Other chemicals produced from aluminum hydroxide in relatively smaller quantities include aluminum phosphate, used as a binder in refractories; sodium aluminum phosphate, a baking powder ingredient; aluminum chlorohydrate, an active agent in antiperspirants; aluminum formate and acetate, used in the dying of textiles; and aluminum stearate and resinate, components of paints, varnishes, water-proofing compounds, and lubricating greases.

Cosmetics and Pharmaceuticals

An important cosmetic application of aluminum hydroxide is in toothpaste. Both finely precipitated (approximately 1-μm median size) and finely ground (less than 7 μm) hydroxides are used in toothpastes. The mildly abrasive character of the hydroxide cleans and polishes teeth, and its chemical inertness makes it easily compatible with other ingredients of the toothpaste formula. Fine-particle hydroxide has been used in some cosmetic powders as a base and filler. Gelatinous aluminum hydroxide is an ingredient of commonly used antacid formulations.

Glass Industry

Aluminum hydroxide is added directly to the glass melt generally in a proportion of between 2 and 4% of the total batch. Alumina helps fusion, reduces viscosity of molten glass, inhibits corrosion of furnace refractories, facilitates annealing, retards devitrification, increases resistance to chemical attack, and improves mechanical strength and appearance of the finished glass.

In vitreous enamel and glazing compositions for whiteware bodies, aluminum hydroxide improves luster and surface smoothness.

Other Applications

Following is a partial list of some miscellaneous applications of aluminum hydroxides:

1. reinforcing pigment in adhesives and adhesive tapes, surgical tapes, and pressure-sensitive tapes

2. extender for titanium dioxide in interior latex wall coatings
3. flocculent and pitch collector in paper processing
4. surface modifier, anchoring agent, and mordant in the textile industry
5. component of printing inks

Analytical Procedures for Aluminum Hydroxides

The industrial production and applications of aluminum hydroxides require analytical procedures for quality control. This quality control involves both physical and chemical analyses. The alumina industry has developed and standardized many of the analytical procedures. These procedures are readily available from the major producers of aluminum hydroxides. The following is a brief summary of some of the methods commonly used by the industry.

Phase Analysis. Although loss of weight on heating (110–1200 °C) can differentiate between pure trihydroxides (34.5%) and oxide–hydroxides (15%), it is not very useful when several phases are present together. Loss of weight on heating also cannot distinguish between the different trihydroxides and oxide-hydroxides.

X-ray powder diffraction is the most effective method for identifying and roughly quantifying the phase composition of a product and is commonly used. Difficulties arise, however, in interpretation of diffraction patterns in the case of poor crystallinity and amorphous products. For example, problems faced in the X-ray analysis and characterization of a commercially produced bayerite (Alcoa C-37 hydrate) product for controlling the production process are discussed by Kramer (37).

Particle Size. Particle size plays an important role in many applications of aluminum hydroxides. A large number of techniques, often related to end use of the product, are employed to characterize particle size distribution of aluminum hydroxide powders.

Sieve analysis, both dry and wet, is commonly used in the 20- (microsieves) to 200-μm size range. Sizes in the 2–200-μm range have been determined by light microscopy, sedimentation (e.g., Andreasen pipet, sedimentation balances, etc.), Coulter counter, and the recently developed Microtrac (Leeds and Northrup) instrument that employs the light-scattering principle. The only reliable procedure for characterizing particles less than 2 μm is by electron microscopy. The SEM proved useful in obtaining information on the particle shape of small hydroxide particles and relating this shape to their behavior in many applications.

4. INDUSTRIAL APPLICATIONS OF ALUMINUM HYDROXIDES

A simple sedimentation test, the rate of sedimentation of a water slurry, is often used in the production plant for routine control of particle size.

DTA, TGA, and DSC Analyses. These techniques are useful for understanding the thermal behavior of aluminum hydroxides, which is particularly important in filler-type applications. Several commercial instruments are available for these analyses; the Du Pont instrument has found wide acceptance in the industry. Thermal analysis often complements X-ray diffraction data in providing information on phase composition.

Alkalinity (Soluble Soda) Determination. Aluminum hydroxide originating from the Bayer process invariably contains alkali. The surface alkalinity (or so-called "soluble soda") is generally quickly determined by making a fixed weight percent slurry in water and determining the alkalinity of the solution by pH determination or acid (HCl) titration. Sodium ion-sensitive electrodes have been investigated.

Chemical Analysis. Impurities commonly analyzed for a majority of applications include Na_2O (total), Fe_2O_3, and SiO_2. Methods of chemical analyses are normally those developed and used by the alumina industry. The hydroxide is first dissolved in boiling concentrated HCl and then analyzed.

Soda (Na_2O) is determined by flame photometry or atomic absorption methods. Gravimetric determination utilizes precipitation of sodium with zinc uranyl acetate reagent. Iron and silicon are determined by atomic absorption or photometric methods. In the case of iron, o-phenanthroline reagent is used to develop a red color after reduction to the Fe^{2+} state by hydroxylamine hydrochloride. Absorbency of the solution is measured at a 510-nm wavelength and corrected for blank measurement. Silica is photometrically determined by the molybdenum blue procedure. In this case, the hydrate is dissolved in nitric acid. Reduction with 2-amino-2-naphthol-4-sulfonic acid and $Na_2S_2O_5$ reagents is followed by addition of ammonium molybdate to develop the characteristic blue color. Absorbance is measured at an 810-nm wavelength and compared with those of standards after correction for blank measurement.

Other Measurements. Other end use related tests include free moisture content, rate of dissolution, and amount of undissolved residue in acids and alkalis, resin and plasticizer absorption, whiteness, slurry viscosity, and specific surface area. Test procedures for these properties are generally selected by taking into account consumer requirements and specifications and are mutually agreed upon between the supplier and user of the product.

Literature Cited

1. Sobolev, I.; Woycheshin, E. A. *Abstracts of Papers*, 172nd National Meeting of the American Chemical Society, San Francisco, CA; American Chemical Society: Washington, DC, 1976; Abstract ORPL 74.
2. Thompson, D. C.; Hagman, J. F.; Mueller, N. N. *Rubber Age (N.Y.)* 1958, Aug.
3a. Woycheshin, E. A.; Sobolev, I. In *Handbook of Fillers & Reinforcements for Plastics*; Katz, H. S.; Milewski, J. V., Eds.; Van Nostrand-Reinhold: New York, 1978.
3b. *Fed Regist.* 1970, *35(74)*, 6211.
4. Geppert, G. A.; Woosley, R. D. U.S. Patent 3 874 889, 1975.
5. Conally, W. J.; Thornton, A. M. *Mod. Plast.* 1965, Oct.
6. Ampthor, F. J.; Kroekel, C. H. Presented at the SPI-RP/C Institute 27th Annual Technical Conference, Feb. 1972.
7. Sprow, T. K.; Connolly, W. J.; Kirke, E. J. Presented at the SPI-RP/C Institue 28th Annual Technical Conference, Feb. 1973.
8. Rizzi, M. A. Presented at the SPI-RP/C Institute 27th Annual Technical Conference, Feb. 1972.
9. Lundberg, C. V. *Abstracts of Papers*, 166th National Meeting of the American Chemical Society, Chicago, Aug. 1973; American Chemical Society: Washington, DC, 1973; Abstract ORPL 50.
10. Bonsignore, P. V.; Manhart, J. H. Presented at the SPI-RP/C Institute 29th Annual Conference, Feb. 1974.
11. Bonsignore, P. V.; Hsieh, H. P. Presented at the Fire Retardant Chemical Association Meeting, Houston, March 14, 1978.
12a. Martin, F. J.; Price, K. R. *J. Appl. Polym. Sci.* 1968, *143*, 58.
12b. Annu. Book ASTM Stand. 1977, D2863-70.
13. Stevens, J. J. Presented at the 9th Electrical Insulation Conference of IEEE, Sept. 1969.
14. Patrick, E. T.; McGrary, C. W. Presented at the 24th Annual Technical Conference of SPE, 1966.
15. Bruins, P. F. *Epoxy Resin Technology*; Wiley: New York, 1968.
16. Kumasaka, S., et al. U.S. Patent 3 737 400 to Toyo Rubber Chemical Industrial Corp., 1973.
17. Norman, A. J.; Cobbledick, D. S. U.S. Patent 3 810 851 to General Tire and Rubber Company, 1974.
18. Taylor, R. C., et al. U.S. Patent 3 821 067 to Atlantic Richfield Company, 1974.
19. Kehr, C. L., et al. U.S. Patent 3 897 372 to W. R. Grace, 1975.
20. Bonsignore, P. V. *Cell. Plast.* 1979, May-June, 163-179.
21. Bonsignore, P. V.; Levendusky, T. L. *Fire Flammability* 1977, *8*, 95-114.
22. Burton, P. U.S. Patent 3 741 929 to ITT Corp., 1973.
23. Aishima, I., et al. Japan, Kokai 7459 841 to Asahi Chemical Industrial Company, 1974; *Chem. Abstr. Jpn.* 1974, *81*, 153769l.
24. Walters, R. B. Ger. Offen. 2 439 490 to General Electric Company, 1975.
25. Thomas, J. L. U.S. Patent 3 826 777 to FMC Corp., 1974.
26. Talsma, H. U.S. Patent 3 786 041 to Standard Oil Company, 1974.
27. Woycheshin, E. A.; Sobolev, I. *J. Fire Flammability/Fire Retardant Chem. Suppl.* 1975, *2*, 224.
28. Sobolev, I.; Woycheshin, E. A. *J. Fire Flammability/Fire Retardant Chem.* 1974, *1*, 13.
29. Bohn, E. *Allg. Pap.-Rundsch.* 1964, *24*, 1987-1991.
30. Hentschel, H. *Wochenbl. Papierfabr.* 1968, *23/24*, 851-854.
31. Hegenbarth, R. *Papeterie* 1968, *3*, 238-239.
32. Murray, H. H. *Paper Coating Pigments*; Monography, Series No. 30; New York, 1966.

33. Koenig, J. J. *Tappi* **1965**, *48* (*12*), 123A-124A.
34. Martifin; Martinswerk GmbH for Chemical & Metallurgical Production: Bergheim, 1970.
35. Gitzen, W. H. *Alumina as a Ceramic Material*; American Ceramic Society: Columbus, OH, 1970.
36. MacZura, G.; Goodboy, K. P.; Koenig, J. J. *Kirk-Othmer Encyclopedia of Chemical Technology*, 3rd ed.; Wiley: New York, 1978; Vol. 2.
37. Kramer, R. A. *Aluminium (Duesseldorf)* **1974**, *50*, 784-785.
38. Woycheshin, E. A.; Sobolev, I. In *Handbook of Fillers and Reinforcements for Plastics*; Katz, H. S.; Milewski, J. V., Eds.; Van Nostrand-Reinhold: New York, 1978; p 246.

5

Activated Aluminas

Activated aluminas represent another group of technically important alumina chemicals. These compounds cover a wide range of industrial and technical applications. Principal uses are as drying agents, adsorbents, catalysts, and catalyst carriers. These products are obtained by thermal dehydration of different aluminum hydroxides in the 250–800 °C temperature range. Water is driven out upon heating of the hydroxides; a highly porous structure of aluminum oxide having a high surface area remains. The physical and chemical natures of the starting hydroxide and the thermal history of dehydration influence the properties of the final "active" product. A wide range of products is obtained by controlling the hydroxide preparation and dehydration processes. The physical properties of the activated alumina, its porous structure, and the chemical nature of the surface area created by the dehydration process are important factors that decide the areas of application for these products. Although the first preparation and uses of activated alumina go back more than 70 years, the material continues to attract considerable scientific interest even today.

Thermal Dehydration of Aluminum Hydroxides and Formation of Transition Aluminas

Classification and Nomenclature. In the thermal dehydration of aluminum hydroxides, the final anhydrous form is corundum (α-Al_2O_3), which is formed above 1100 °C. In the intermediate temperature range, various so-called transition aluminas are formed. The variables of the dehydration process and the structures of the transition forms between the stable hydroxides and the final α-Al_2O_3 have been studied extensively over the past 40 years by many groups of investigators all over the world.

The transition aluminas, though fairly reproducible, are neither completely anhydrous nor crystallographically well defined, which is the reason for some confusion regarding their identification and classification.

The designation γ was originally given by Ulrich (1) to an ill-defined alumina, and this term came, subsequently, to be used generically for all transition phases encountered in the low-temperature calcination of aluminum compounds and the oxidation of aluminum. The first systematic investigation of the transition aluminas was carried out at Alcoa, and a classification scheme was proposed by Stumpf et al. (2) in 1950. Later investigations by other workers, notably Papée et al. (3), Day and Hill (4), and Ginsberg et al. (5), confirmed the principal features of the structures determined by Stumpf et al. During a symposium held in West Germany in 1957, an attempt was made to standardize the nomenclature, and the Alcoa designations were generally accepted for classification of the transition aluminas. Additional alumina forms have been reported in the literature, but in most cases these forms lack clear X-ray or structural data, which has prevented their general acceptance. Lippens and Steggerada (6) have suggested the following classifications based on the temperatures at which the aluminas are obtained from the hydroxides:

1. Low-temperature aluminas: $Al_2O_3 \cdot nH_2O$ in which $0 < n < 0.6$. These forms are obtained at dehydration temperatures not exceeding 600 °C. These forms are also called the γ group and include the forms ρ-, χ-, η-, and γ-aluminas and are considered to have a lower order of crystallinity.

2. High-temperature aluminas: nearly anhydrous forms obtained at temperatures between 800 and 1000 °C. These forms are termed the δ group and include the forms κ-, θ-, and δ-aluminas. These forms are relatively better ordered; the result is sharper X-ray diffraction patterns.

Following the work of Stumpf et al., Tertian and Papée (7), and others (4, 5), the X-ray powder diffraction analysis has proved to be the most practical tool for identifying the various transition aluminas. However, interpretation of powder data is not always straightforward. The transition forms represent different stages of reordering of the lattice. Thus, line broadening and the possible presence of several transition forms together introduce complications to interpretation of powder data.

The ρ form is considered to be largely X-ray amorphous and was not included in Stumpf's original classification. This name was proposed by Papée and Tertian for a form obtained by thermal dehydration of gibbsite under high vacuum. This form was later found to be the major constituent of the so-called rehydratable aluminas produced by very rapid dehydration of the trihydroxides.

Characteristic line spacings and relative intensities of X-ray powder diffraction patterns of the various transition aluminas are given in Table 5.I. As can be seen, the differences between the γ and η and between the θ and δ forms are quite small and are easily concealed in the case of mixtures.

Dehydration Sequences. TRIHYDROXIDES. The dehydration sequence for gibbsite in air differs from that of bayerite and nordstrandite. Also, a particle size effect exists: in larger gibbsite particles, about 25% boehmite can

Table 5.I.
X-Ray Powder Diffraction Patterns for Transition Aluminas

χ		γ		η		κ		δ		θ	
d^a	I/I_o^b	d	I/I_o	d	I/I_o	d	I/I_o	d	I/I_o	d	I/I_o
2.398	50	2.755	20	4.600	30	6.102	10	5.040	40	5.472	5
2.290	10	2.430	30	2.755	20	4.484	5	4.570	60	5.063	5
2.125	30	2.292	30	2.400	50	3.048	30	4.070	70	4.570	10
								2.866	20	4.020	5
1.984	20	2.121	6	2.298	30	2.814	50	2.738	50	2.859	60
1.535	10	1.993	90	1.985	100	2.730	5	2.592	10	2.731	100
1.397	100	1.965	40	1.519	20	2.585	100	2.455	40	2.462	80
—	—	1.519	20	1.402	100	2.424	20	2.286	20	2.314	40
—	—	1.397	100	1.210	10	2.327	35	1.993	30	2.264	30
—	—	—	—	—	—	—	—	1.953	10	2.028	80
—	—	1.141	10	1.141	10	2.166	5	1.914	10	1.193	40
—	—	—	—	—	—	2.127	90	1.807	10	1.800	5
—	—	—	—	—	—	2.067	5	1.542	45	1.538	20
—	—	—	—	—	—	1.993	15	1.514	40	1.484	20
—	—	—	—	—	—	1.953	5	1.497	40	1.456	20
—	—	—	—	—	—	1.877	30	1.455	80	1.429	10
—	—	—	—	—	—	1.831	5	1.425	60	1.407	30
—	—	—	—	—	—	1.639	20	1.393	100	1.392	100
—	—	—	—	—	—	1.488	5	1.296	40	—	—
—	—	—	—	—	—	1.453	5	1.260	20	—	—
—	—	—	—	—	—	1.437	50	1.141	40	—	—
—	—	—	—	—	—	1.395	95	—	—	—	—
—	—	—	—	—	—	1.344	5	—	—	—	—

NOTE: Filtered Cu Kα radiation was used.
^{a}d denotes line spacings.
$^b I/I_o$ denotes relative intensities.

form within the crystal under hydrothermal conditions due to the buildup of water vapor pressure within the crystals on heating. Relatively little boehmite is formed in fine particle size hydroxides. Boehmite formation is also quite less (less than 5%) with bayerite and nordstrandite due to the small size of the crystals.

If the particle size of the starting gibbsite is small (≤ 10 μm) and the water vapor pressure in the atmosphere in which the particles are heated is low, then the nearly X-ray-indifferent χ-Al_2O_3 is formed on heating the gibbsite between 250 and 350 °C. The χ form has been shown to possess a highly disordered hexagonal layer structure similar to that of the original pseudohexagonal gibbsite (8). The stacking sequence of the layers is strongly disordered. On further heating, the nearly anhydrous κ form, which also has a hexagonal layer lattice (9), is formed above 800 °C. Conversion of the κ form to corundum occurs at about 1100 °C. The structure of χ-alumina contains appreciable amounts of hydroxyl ions. With the loss of practically all OH ions, the anion lattice of the κ form is better ordered than that of the low-temperature χ form.

The dehydration bath of both bayerite and nordstrandite is different from that of gibbsite. The low-temperature transition form in this case is η-alumina, formed at a temperature of 200–350 °C. This form converts to the θ form at 850 °C and finally to α-Al_2O_3, corundum, above 1000 °C. These transition forms are reported to have a spinel-like structure that is tetragonally deformed (9). θ-Alumina is isostructural with β-Ga_2O_3, which has been thoroughly studied due to the availability of large single crystals. θ-Alumina's structure can be described as a deformed spinel type with aluminum ions occupying predominantly tetrahedral positions (10).

Fast dehydration of all the trihydroxides (under vacuum or rapid exposure to high temperatures) results in the formation of the amorphous ρ-alumina, which changes to η-alumina and further to θ-alumina in the sequence

$$\text{trihydroxide (all forms)} \xrightarrow{\text{fast dehydration}} \rho\text{-}Al_2O_3 \xrightarrow{400\ °C} \eta\text{-}Al_2O_3 \xrightarrow{750\ °C} \theta\text{-}Al_2O_3 \xrightarrow{1200\ °C} \alpha\text{-}Al_2O_3$$

ρ-Alumina rehydrates with water to form bayerite irrespective of the original form of the starting trihydroxide. This property has been utilized for the preparation of bayerite and activated alumina products.

The transition temperatures shown in the above sequences are only approximate. Kinetic factors influence the various transformations to a large degree. In addition to the water vapor pressure, the presence of alkali and other impurities (including additives) also affects transformation temperatures and products. Stumpf et al. have observed that the same transition forms are encountered for both short and long heating periods; the effect of long holding at a temperature is merely to lower the transformation temperatures. The long-period specimens are notable for a lack of the amorphous material that coexisted with the short-period decomposition products in varying amounts. The effect of steam is complex. In general, steam increases the decomposition temperature of the hydroxides, decreases the decomposition temperature of the low-temperature transition aluminas, and has no effect on the transformations of the nearly anhydrous high-temperature forms. Transformation temperatures during activation in humid air tend to be intermediate between those for dry air and steam activation.

OXIDE–HYDROXIDES. γ-Al_2O_3 is formed on heating well-crystallized boehmite to 400–500 °C. The γ form has been shown to possess a tetragonally distorted defect spinel structure (11). γ-Al_2O_3 usually contains up to 0.2%

H_2O. Recently Leonard et al. (12) have suggested that the η and γ forms should be distinguished from the standpoint of their oxygen packings. On the basis of a study of radial electron distributions, Leonard et al. calculated that the average Al–O bond length in γ-Al_2O_3 (1.820–1.818 Å) is smaller than that in η-Al_2O_3 (1.838–1.825 Å). Consequently, the oxygen lattice is more densely packed in γ-Al_2O_3 than in η-Al_2O_3; this observation is certainly related to the higher density of stacking faults in η-Al_2O_3. The distribution of cations is also slightly different in the two cases. Although octahedral sites are preferentially occupied, the fraction of cations in tetrahedral position is slightly higher in γ-Al_2O_3 than in η-Al_2O_3. Thus, the main structural differences between the two modifications are found in the lower packing density of the oxygen lattice in η-Al_2O_3 and the slightly higher occupation of tetrahedral positions in γ-Al_2O_3.

The nearly water free δ-oxide occurs on further heating to 750–850 °C. Several investigators have reported the appearance of the θ form at around 1000 °C before conversion of the δ form to corundum at about 1100 °C. The structure of δ-alumina has some similarity with that of θ-alumina. A tetragonal unit cell is generally assigned to this form (13). Once again the dehydration process is strongly affected by the presence of impurities and the degree of crystallinity of the boehmite.

Both η- and γ-aluminas are often reported in the product of thermal decomposition of gelatinous (pseudo-) boehmite at 450 °C. The transition sequence is similar to that of bayerite. Considerable line broadening is observed in this product; this line broadening hinders exact identification.

Diaspore decomposes directly to α-Al_2O_3 without passing through any transition stages (14). At temperatures of less than 500 °C, the product exhibits line broadening and a high surface area. Recrystallization to α-Al_2O_3 occurs on prolonged heating and higher temperatures.

$$\text{diaspore} \xrightarrow{500\ °C} \alpha\text{-}Al_2O_3$$

Because of the structural similarity between the hexagonal closely packed lattices of diaspore and corundum, phase transformation requires relatively small rearrangements of the oxygen and aluminum positions. A summary of the different alumina thermal transformations is shown in Figure 5.1.

Dehydration Mechanisms. Few chemical processes have been studied as thoroughly as the thermal dehydration of aluminum hydroxides. The structural transition of the hydroxides to the stable α-Al_2O_3 form follows a common rearrangement pattern. In the first step, hydrogen bonds between adjacent layers are eliminated. In the simplest case, that of diaspore, whose structure is very similar to that of α-Al_2O_3, the subsequent nucleation of α-

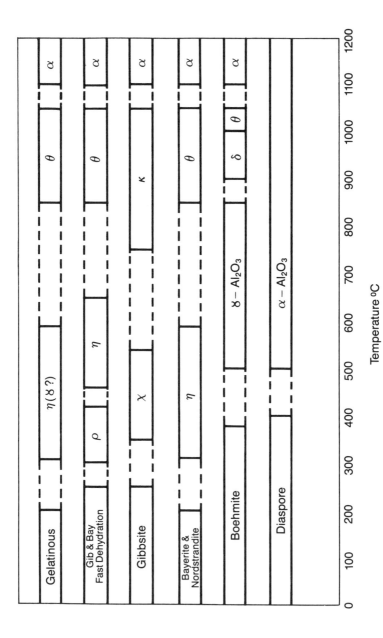

Figure 5.1. Decomposition sequence of aluminum hydroxides.

Al_2O_3 requires only minor structural rearrangement. Epitaxial growth of corundum on diaspore has been demonstrated (15). The conversion of diaspore to corundum takes place in situ (topotactic): the a, b, and c axes of the oxide-hydroxide become the c, a, and $(1\bar{1}0)$ direction of α-Al_2O_3 (16, 17). Evidence (18) that an intermediate structure occurs during transformation, represented by an alternating sequence of areas with ordered and disordered cation distribution, exists.

A model for the dehydration of boehmite to γ-Al_2O_3 has been recently proposed by Wilson (19). In this model the development of a porous microstructure is considered to be an integral part of the reaction mechanism. The pseudomorphosis relationships involved in the topotactic dehydration of boehmite to γ-Al_2O_3 are as follows (11, 20):

for a_{BO}

$$3.700 \text{ Å} \longrightarrow (001)_\gamma$$

for b_{BO}

$$12.227 \text{ Å} \longrightarrow (110)_\gamma$$

for c_{BO}

$$2.868 \text{ Å} \longrightarrow (1\bar{1}0)_\gamma$$

The stoichiometry of the reaction

$$2AlOOH \longrightarrow Al_2O_3 + H_2O$$

indicates that one-quarter of the boehmite oxygen lattice is removed during the decomposition. A mechanism proposed previously (20, 21) is one that involves elimination of H_2O by an internal condensation of protons and hydroxyl groups between the layers followed by collapse of the layered structure. In the case of boehmite, the initial product of such a collapse mechanism would result in the formation of the cubic closely packed framework of the spinel structure (Figure 5.2). However, the dehydration of boehmite via a collapse mechanism does not require the formation of pores. As has been suggested for other layered hydroxides, strains set up in the crystal during collapse might produce fragmentation, which results in an apparent porous structure and slight misorientations between the adjacent oxide blocks thus formed. Wilson has given evidence that the regularity of the orientation and spacing of the lamellar porous microstructure, the coherence of the γ-Al_2O_3 electron diffraction patterns, and the coherence of the $(111)_\gamma$ lattice fringes across the pores can be interpreted to mean that the formation of the pore

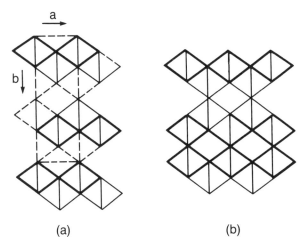

Figure 5.2. Idealized structures of (a) boehmite and (b) collapsed boehmite. Reproduced with permission from Reference 19. Copyright 1979 Academic.

system is crystallographic and a necessary feature of the dehydration mechanism. A mechanism involving countermigration of protons and Al cations similar to that suggested for $Mg(OH)_2$ by Ball and Taylor (22) but with the added constraint that the direction of diffusion should be governed by the hydrogen-bond chain in the boehmite structure may be an explanation. This mechanism is illustrated in Figure 5.3. Protons are lost from between the boehmite layers by a diffusion process along the hydrogen-bond chains, and the interlayer sites thus created are occupied by counterdiffusing Al cations. The elimination of H_2O is due to the abstraction by the protons of oxygen anions and hydroxyl groups from the region of the now-vacant Al sites. This elimination results in the formation of regions of void that appear as pores, the surfaces of which will presumably be highly protonated. The formation of the spinel lattice of γ-Al_2O_3 must be accomplished by final reorganization of the Al cation positions. This mechanism produces a coherent skeleton of γ-Al_2O_3 based on the original boehmite lattice and intersected by pores perpendicular to the direction of diffusion. This skeleton is claimed by Wilson to be consistent with the observed fine lamellar (001) pore system and the coherence of γ-Al_2O_3 electron diffraction patterns and lattice fringes. A small degree of relative movement of the boehmite layers is necessary for the formation of octahedral Al cation sites and the cubic closely packed oxygen lattice of spinel structure. This movement is approximately $C_{BO}/2$ (1.43 Å), which is much less than that (2.70 Å) required for the collapse mechanism; thus, the countermigration theory is additionally supported.

In a subsequent study of the $\gamma \rightarrow \delta \rightarrow \theta \rightarrow \alpha$ transformation at increased temperatures, Wilson and McConnell (23) report on the gradual nature of

5. ACTIVATED ALUMINAS

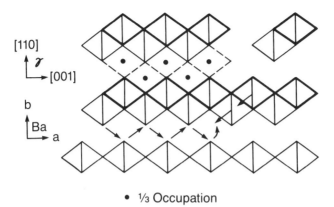

* ⅓ Occupation

Figure 5.3. Schematic illustration of the proposed mechanism for the dehydration of boehmite. Reproduced with permission from Reference 19. Copyright 1979 Academic.

the γ → δ transformation found from the continuous variations of spinel subcell parameters and from the gradual appearance of superstructure reflections in X-ray diffraction patterns. Thus, no distinct dividing line between γ- and δ-Al_2O_3 can be found. These patterns may be considered to represent two different structure types: γ-Al_2O_3 is a defect spinel in which the vacancies are distributed randomly, whereas in δ-Al_2O_3 the vacancies are ordered on octahedral sites; this gives a spinel superstructure. The θ structure can form by rearrangement of the cations in an approximately cubic, closely packed oxygen array.

The structure of the boehmite precursor and the proposed mechanism of its dehydration impose a cubic, closely packed oxygen lattice on the system, and thus the sequence γ-Al_2O_3 → δ-Al_2O_3 → θ-Al_2O_3, all based on this type of anion stacking, provides a kinetically favored reaction route. The gross reorganization necessary for the topotactic nucleation of hexagonally closely packed α-Al_2O_3 only becomes kinetically feasible at high temperatures.

In the decomposition of the trihydroxide, the first step in the reaction sequence is the diffusion of protons to adjacent OH groups and the subsequent formation of water, which begins at about 200 °C (24). This process removes the binding forces between the layers of gibbsite structure. Separation of the layers and a distortion of the arrangement of the original OH octahedra are the results. Also, if the water formed cannot rapidly diffuse out of the larger trihydroxide particles, hydrothermal conditions may develop locally; formation of boehmite results. The separated layers break up into numerous areas of short-range order corresponding to the diffuse χ-alumina form. The complete absence of ordering resulting from very rapid dehydration is represented by the amorphous ρ-alumina form.

A detailed mechanism of the above processes has been proposed by Rouquerol et al. (25, 26) from a study of the thermal decomposition of gibbsite under low pressures. Rouquerol et al. postulated a model in which the first step involves the incomplete transformation of gibbsite into boehmite in the temperature range of 180–205 °C. Boehmite is formed in the interior of the crystal. Water is released by mutual displacement of hydroxyl ions via structural channels or cleavage planes in the gibbsite structure. The slowness of this migration and difficult desorption from the end openings of the channels or from the edges of the cleavage planes help to maintain hydrothermal conditions within the crystal. The presence of higher density boehmite (3.02 g cm^{-3}) inside the lower-density gibbsite (2.4 g cm^{-3}) results in gaps that are filled with water. The presence of this occluded water has been detected by nuclear magnetic resonance studies. The formation of boehmite ceases when the intercrystalline water pressure is lowered due to formation of cracks and possibly thinning down of the gibbsite shell, which could result from the advance of the boehmite phase itself or surface decomposition of gibbsite according to the second step of the process. The transition from the first (boehmite formation) step to the second step (formation of transition alumina) depends on kinetic factors such as texture change and temperature increase. The second step consists of the "drilling" of micropores into the gibbsite with the reaction plane advancing parallel to the basal plane of the crystal. This process involves two phenomena. The first one is the dehydroxylation of the "structural channels" lined only with hydroxyl groups. The second phenomenon is related to the rate of desorption, which controls the size of the micropores and involves some countermigration of Al ions as in the case of boehmite.

With increasing temperature, reordering and densification proceed within the layers of the former gibbsite lattice. X-ray diffraction patterns of a sample heated above 800 °C indicate a better ordered transition form—κ-alumina. The temperature range in which this form occurs largely depends on the prehistory of the sample, especially its content of alkali metal oxides (5). κ-Alumina has a water content of about 2%, which is removed only after heating above 1100 °C for nearly 1 h. This observation suggests some stabilizing influence of OH ions on the structure of transition aluminas. Heating above 1100 °C results in the formation of a three-dimensional network of corundum crystallites in contrast to the layer structure of transition forms inherited from the precursor gibbsite lattice. This last step of the transition sequence comprises further densification of anion lattice from cubic to hexagonal closely packed arrangement as well as a change in aluminum positions in tetrahedral interstices to the octahedral arrangement of α-Al$_2$O$_3$.

Rehydration. Rehydration of the amorphous ρ-alumina to bayerite was mentioned previously. Commercial experience with products obtained by rapid dehydration of Bayer gibbsite shows that these products rehydrate to

5. ACTIVATED ALUMINAS   83

pseudo-boehmite and then to bayerite in contact with water. The appearance of pseudo-boehmite is quite rapid, and the rate increases with temperature. For example, more than 60% of the product was converted to pseudo-boehmite in about 4 h at 90 °C. Recrystallization to bayerite takes place more slowly; the rate is highest in the temperature range of 45–65 °C. Little bayerite is formed above 90 °C even after long aging periods. The rehydration property of ρ-alumina has been a key factor in the development of commercially important active alumina adsorbent and catalyst products. This fact may account for the proliferation of patent literature in this area.

In general, rehydration of γ- and η-aluminas is quite slow under room-temperature conditions. In desiccant uses of activated aluminas, the moisture-removal capacity shows a gradual decline with increased usage and number of regeneration cycles. This decline has been attributed to slow rehydration, which results in the formation of aluminum hydroxides. The γ- and η-aluminas convert to boehmite under hydrothermal conditions; nearly total conversion takes place in about 6 h at 250 °C.

Thermal Effects

Differential thermal analysis (DTA) and thermogravimetric analysis (TGA) for gibbsite and boehmite are shown in Figure 5.4. In the case of gibbsite, most of the water (nearly 30% by weight from a total of 34.6%) is driven off at temperatures below 400 °C. Two endothermic peaks below this temperature appear in the DTA curve for a coarse-grained gibbsite. The first appears at around 230 °C. This peak is followed by an exothermic effect at approximately 280 °C. This effect is attributed to the formation of intergranular hydrothermal boehmite due to buildup of water vapor pressure within the gibbsite particle. This exothermic effect is not observed in the DTA curve of fine gibbsite. The endothermic effect at 550 °C caused by the decomposition of the boehmite in the coarse-grained product is also missing in the case of fine gibbsite. The endothermic effect at 550 °C is clearly seen in the DTA curve for well-crystallized boehmite.

De Boer (27) has plotted a thermal dehydration curve (Figure 5.5) for gibbsite showing residual water content as a function of temperature and has delineated the three steps in which boehmite, χ-Al_2O_3, and γ-Al_2O_3 (from boehmite) are formed. The three steps are more clearly displayed in the differential curve.

The DTA and TGA curves of coarse- and fine-grained bayerite are nearly identical to those of gibbsite. Some boehmite is also formed due to intergranular hydrothermal conditions in large particles of bayerite. Finely divided material transforms between 230 and 350 °C to η-Al_2O_3.

Eyraud and co-workers (28–30) measured the electrical energy input required for dehydration of gibbsite at a constant heating rate of 4.5 °C/min. The following heat effects were observed: an endothermicity with a maxi-

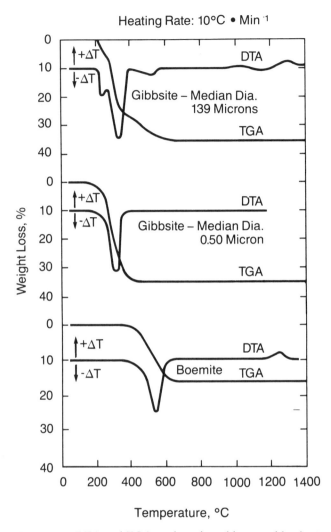

Figure 5.4. DTA and TGA analysis for gibbsite and boehmite.

mum at 240 °C, a much greater endothermicity with a maximum at 327 °C, a slight exothermic effect between 350 and 400 °C, and a final small endothermic effect with a maximum at 500 °C. Adding all these gives the heat requirement for dehydration of 73.7 kcal/mol ($Al_2O_3 \cdot 3H_2O$). By combining these measurements with weight loss calculations, these workers determined thermal requirements for each step of dehydration:

$Al_2O_3 \cdot 3H_2O$ to $Al_2O_3 \cdot 2.8H_2O$ = 3.5 kcal/g of H_2O

$Al_2O_3 \cdot 2.8H_2O$ to $Al_2O_3 \cdot 2.7H_2O$ = 0.9 kcal/g of H_2O

$Al_2O_3 \cdot 2.7H_2O$ to $Al_2O_3 \cdot 0.5H_2O$ = 1.1 kcal/g of H_2O

$Al_2O_3 \cdot 0.5H_2O$ to Al_2O_3 = 2.0 kcal/g of H_2O

Jackson and Jones (31) have reviewed various values quoted in the literature for the magnitude of the endotherm of dehydration of gibbsite and report their own determinations using a differential scanning calorimetric (DSC) method. They suggest that the heat of dehydration of gibbsite (to

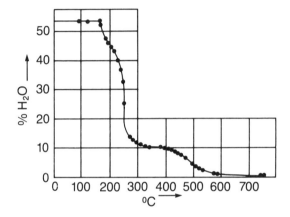

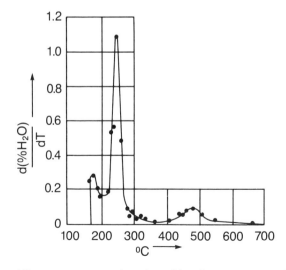

Figure 5.5. Water content as a function of heating temperature for gibbsite (percent H_2O is on an Al_2O_3 basis). Reproduced with permission from Reference 26. Copyright 1979 Academic.

water in the vapor state) is 43 kcal (mol of $Al_2O_3 \cdot 3H_2O)^{-1}$ [280 cal (g of $Al_2O_3 \cdot 3H_2O)^{-1}$].

Goton (32) reported heating of well-crystallized boehmite at a rate of 4.5 °C min^{-1} to produce a strong endothermic effect at 510 °C. The total heat required for dehydroxylation was 35 kcal mol^{-1}. As can be seen in the DTA analyses, the exothermic effect at around 1200 °C is related to the transformation of the high-temperature transition form to α-Al_2O_3. Yokokawa and Kleppa (33a) determined the enthalpies of the following transformations:

$$\gamma \rightarrow \alpha = -5.3 \text{ kcal mol}^{-1}$$

$$\kappa \rightarrow \alpha = -3.6 \text{ kcal mol}^{-1}$$

$$\delta \rightarrow \alpha = -2.7 \text{ kcal mol}^{-1}$$

These values have been disputed in other investigations (34). Sabatier has reported the heat of decomposition of diaspore to α-Al_2O_3 as 19.9 kcal mol^{-1} (35).

Textural Changes: Development of Pore Structure and Surface Area

Texture Development. Heating of aluminum hydroxides results in dehydroxylation and is also accompanied by a change in density. The formation of extensive pores and cracks and a large increase in internal surface area occur. Although the dehydration process is considered pseudomorphous, involving little change in the size or shape of the particles, some swelling of the gibbsite particle occurs until 200 °C followed by a 2.5% linear contraction at higher temperatures as shown by dilatometric studies (33b).

Figure 5.6 shows the emergence of crevices and cracks in heated gibbsite crystals. The internal surface area produced by the disintegration of the gibbsite structure reaches a maximum of about 350–400 m^2 g^{-1} in the temperature range of 300–400 °C and decreases thereafter.

Russell and Cochran (36) observed that the surface area accessible to nitrogen remained below 2 m^2 g^{-1} until the weight loss was about 8%. Thereafter, the surface area rose to a peak of about 350 m^2 g^{-1} for 1-h calcination at 400 °C in dry air. Blanchin (37) pointed out that the highest surface area corresponds to the composition $Al_2O_3 \cdot 0.5H_2O$. The relationship between heating variables and the specific surface area of heated gibbsite has been thoroughly investigated. Although the absolute surface area value depends on several factors such as rate of heating, water vapor pressure, particle size, and purity of the starting material, the common behavior is a sharp increase in internal surface area up to a temperature of about 400 °C followed by a gradual decrease between 400 and 1200 °C. Typical results for a

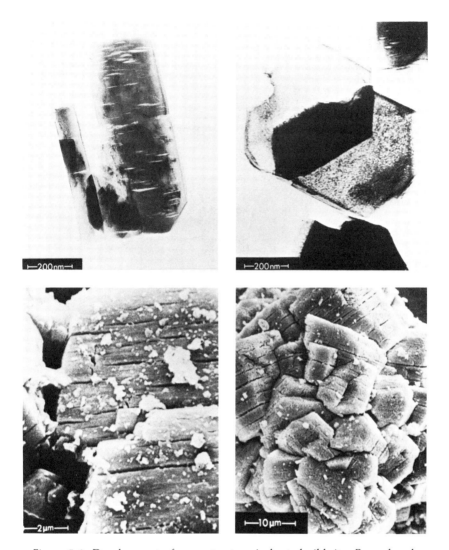

Figure 5.6. Development of pore structure in heated gibbsite. Reproduced with permission from Reference 51. Copyright 1969, BACO.

Bayer hydroxide (Figure 5.7) shows the relationship between temperature, specific surface area (BET nitrogen adsorption method), density, and weight loss on ignition during the course of thermal decomposition.

The geometry (shape and size distribution) of the pore structure formed by gibbsite dehydration has been studied extensively (*38–40*).

Nitrogen adsorption isotherms of the dehydration products have been used to learn about their porous structure. The crystalline trihydroxides have a low surface area and small pore volume; the shape of the isotherms is

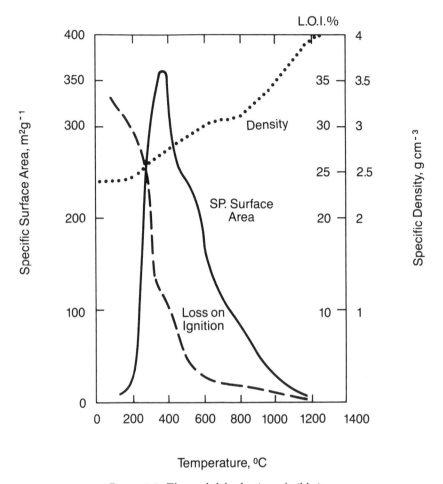

Figure 5.7. Thermal dehydration of gibbsite.

in agreement with a picture of loosely packed, nonporous particles with sizes of several micrometers. When the trihydroxides are heated to just below the temperature at which low-temperature transition forms are produced (200–220 °C), the adsorption isotherm shows that pores are formed. The hysteresis loop in the adsorption curve has been interpreted (41) to visualize pore shape. In the case of gibbsite, these pores have been described as "ink bottle" type or "funnel" type pores, with a large volume but narrow openings. In the case of bayerite and nordstrandite, the pores are more slit-shaped in character.

When the material is heated above the decomposition temperature (230–250 °C), nitrogen adsorption on the material at low relative pressures increases rapidly; this increase corresponds to a similar rise in the BET surface area to more than 350 m² g⁻¹. That most of the adsorbed nitrogen is present in a micropore system and only a small part of the surface area is present in pores wider than about 20 Å can be inferred from the adsorption–desorption isotherm. Lippens (42) examined the micropore volume by studying the adsorption of lauric acid from pentane solution. The picture that emerges from this study is a system of pores of about 20-Å width separated by solid lamellae of about 30-Å thickness.

When the material is heated further, the surface area decreases. A considerable change in the shape of the adsorption isotherm occurs in the product heated to above 550 °C. The BET surface area drops to about half its highest value. Calculation of surface area and pore volume from the isotherms shows that the micropore volume has decreased to a very low value. The remaining pore volume is accounted for by larger rod-shaped macropores of 50–300-Å size.

The optical characteristics of the hydroxide crystal during heating are indicative of textural changes. When a single crystal of gibbsite is heated, the crystal loses its transparency; the resulting product is initially birefringent (6) due to the pronounced texture of its pore system. The nature of birefringence, which can be determined under a polarizing microscope, changes with increasing temperature. At low temperature the birefringence is due to a system of pores parallel to the cleavage plane of the gibbsite lattice. These pores constitute the micropore system and disappear gradually by sintering at higher temperatures. Orientation of rodlike pores then becomes dominant in determining optical behavior.

The two crystalline oxide–hydroxides, boehmite and diaspore, behave similarly on dehydration with regard to the evolution of surface area and pore system. Decomposition starts at about 400 °C; the evolution of BET surface area is less dramatic; the area reaches a maximum value of 90–100 m² g⁻¹ at about 500 °C. The surface area is accounted for almost entirely by the micropore volume. Above about 550 °C, the micropore volume disappears with a corresponding decrease in BET surface area to 15–20 m² g⁻¹. However, unlike the changes in the decomposition of trihydroxides, the nitrogen isotherms do not show any appreciable change in types of pores. Even after well-crystallized boehmite is heated to 750 °C, the pore system still retains the character of slit-shaped pores between plane-parallel layers.

In general, the original (before dehydration) surface area of gelatinous aluminum hydroxide decreases with increasing crystallinity. However, dehydration produces a smaller increase in surface area the less crystalline the material. In extreme cases of highly amorphous products, a decrease in surface area is observed. This decrease is shown in Table 5.II. This loss in surface area is due, almost completely, to loss of water, which causes a shrinkage of particles. No new internal pores are formed, nor do the particles sinter together to any appreciable extent. In crystalline or partly crystalline products, the quantity of new pores produced by dehydration depends on the

Table 5.II.
Surface Area of Hydroxides of Different Crystallinity and of Their Dehydration Products

Type of Boehmite	Surface Area (m^2/g)		
	Dried at 120 °C	Dehydrated at 500 °C	Change
Well crystallized	1.3	65.3	+64
Microcrystalline	64	100	+36
Microcrystalline	68	99	+31
Microcrystalline	100	101	+1
Microcrystalline	201	180	−21
Microcrystalline	255	208	−47
Gelatinous	395	257	−138
Gelatinous	490	316	−174
Gelatinous	609	398	−211

SOURCE: Reproduced with permission from Reference 6. Copyright 1970, Academic Press.

crystallinity of the original hydroxide. The adsorption isotherms confirm this picture.

Knowledge of pore volume and structure has also been obtained by the mercury penetration technique. Mercury porosimetry is based on the capillary law governing liquid penetration into small pores. This law, when applied to a nonwetting liquid such as mercury filling a cylindrical pore, is expressed as

$$D = -(1/p)4\sigma \cos \theta$$

where D is the pore diameter, p is the applied pressure, σ is the surface tension, and θ is the contact angle. The volume of mercury, V, penetrating the pores is measured as a function of applied pressure. This p–V information serves as a unique characterization of pore structure. Although the above relationship applies strictly to the cylindrical shape, which does not represent true pore geometry in most materials, the procedure is generally accepted as a means for characterizing pore size distribution. The practical uses of mercury porosimetry in the study of porous solids with examples of studies on alumina are presented by Moscou and Lub (43). Some examples of pore volume distributions measured for commercial activated aluminas (Alcoa S-100, H-151, and F-1) are presented in Figure 5.8.

Modification of Pore Structure. The pore structure of activated aluminas plays an important role in their application as adsorbents and catalysts. Many methods have been proposed and practiced to control and modify the pore structure to meet specific application requirements. These include the following.

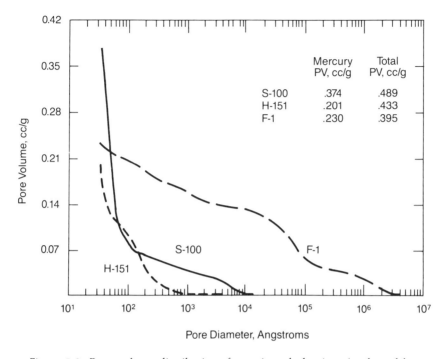

Figure 5.8. Pore volume distributions for activated aluminas (evaluated by mercury penetration, effective for pores down to 35 Å).

TYPE AND METHOD OF PREPARATION OF THE HYDROXIDE. The influence of the structure of the original hydroxide on the evolved pore structure has been discussed earlier. This information affords a means, within certain limits, of obtaining a product with a predictable and reproducible pore size distribution.

The preparation method of the hydroxide also can influence the pore structure significantly, and a large amount of patent literature has accumulated over the years describing preparation techniques that result in the evolution of some desired pore geometry and size distribution. A recent publication by Kotanigawa et al. (40) analyzes the products from 11 different methods of preparation, extending the earlier work of Lippens (42). Preparation methods included are neutralization of aluminum salts (sulfate, nitrate, and chloride) with ammonia and urea, precipitation from sodium aluminate by CO_2 gassing, and thermal decomposition of organic and inorganic aluminum salts (isopropoxide and chloride). This research reports that aluminas prepared by using aqueous ammonia as the precipitating agent exhibited a single pore size distribution of about 20-Å radius. Similar structures were found for products from CO_2 neutralization of sodium aluminate and pyrolysis of aluminum nitrate. On the other hand, products from urea neutralization and aluminum isopropoxide decomposition exhibited a bimodal pore size distribution with radii of about 20 and 50 Å. Evidence from adsorption and desorption experiments and from SEM observations were analyzed to conclude that the unimodal distribution at 20 Å is composed of tubular or ink bottle-shaped pores formed between individual cubic particles or cubic particles built in layers with platelike particles. For the bimodal distribution, probably the smaller 20-Å pores are slit-shaped, formed as spaces between plate structures, whereas the bigger 50-Å pores are tubular or ink bottle shaped, formed as spaces between cubic particles that are composites of layered plate structures.

INCORPORATION OF BURNOUT MATERIALS. An organic material (e.g., polymers, carbon, cellulose, sawdust, etc.) is incorporated into the hydroxide during its preparation stage. This material is allowed to burn off during thermal dehydration; this burn-off produces an additional pore structure. The properties and quantity of the burnout material influence the final pore structure of the product (44, 45).

The pores formed by the burnouts are usually some orders of magnitude larger than the pores formed by dehydroxylation. These large pores serve a useful function by providing a passage through which the smaller micropores become accessible to gas molecules.

A localized temperature rise during combustion of the burnout material can alter the properties of the material immediately surrounding the burnout. This phenomenon imposes some restrictions on the use of this technique. Problems also arise in controlling the rate of burning of the burnout during activation; loss of control of the burning rate results in poor reproducibility of product characteristics.

SUBLIMATION. Riekert and Weber (46) have reported the use of sublimation to generate macropore volume during the dehydration of aluminum hydroxides. These researchers have suggested the use of melamine (triaminotriazine) as the subliming material. Melamine has been shown to be completely sublimed on heating to 500 °C; a large macropore volume in the dehydrated product remains. The sublimation procedure avoids some of the problems of the burnout method such as the effect of heat produced during combustion and lack of control over the burning rate.

FORMING PROCEDURE. Agglomeration, extrusion, and pelletizing are common methods for forming alumina adsorbents and catalysts. The influence of forming variables such as liquid content, binder use, mechanical factors (pressure and rotation speed), and others also can be varied to alter the pore (principally macropore) volume and structure of the final product. Time–temperature conditions maintained during the aging of the formed products prior to final activation have been claimed in various patent disclosures to affect volume fraction porosity and pore size distribution.

Other Sources of Transition Aluminas

Some transition aluminas are obtained by methods other than thermal decomposition of aluminum hydroxides. Two common sources are oxidation of aluminum metal and thermal decomposition of aluminum salts such as nitrate, chloride, and sulfate.

In the presence of oxygen, aluminum is thermodynamically unstable with respect to its oxide, Al_2O_3. However, at low and moderate temperatures the rate of reaction between oxygen and aluminum approaches zero after a short period of rapid oxidation. This characteristic is due to the formation of a dense layer of surface oxide, which inhibits transport of oxygen to the reaction surface. The mechanism of this process has been widely studied and reported (47, 48). The oxide layer so formed is highly amorphous up to a temperature of 400 °C. Above this temperature a structural ordering process takes place; the formation of γ-Al_2O_3 is the result. A mixture of γ- and η-alumina was found on 99.99% pure aluminum after 16 h at 400 °C in an oxygen atmosphere. The presence of water vapor in the oxidizing atmosphere had a rate-increasing effect on the formation of these aluminas. The crystalline ordering and growth of γ-Al_2O_3 disrupts the continuous, amorphous oxide matrix; channels to the base metal surface are created, and thus the oxidation process is accelerated. The γ- and η-aluminas are the primary products of this oxidation process at temperatures above about 400 °C.

The thermal decomposition of aluminum salts—nitrate, chloride, sulfate, acetate, and others—generally yields an amorphous product at low temperatures of less than about 700 °C. Most of these salts are highly hydrated with 6–18 molecules of water. The decomposition product at these temperatures contains considerable amounts of anions. At temperatures above 800 °C, γ-Al_2O_3 lines appear in the X-ray diffraction pattern of the product, which converts to δ-, θ-, and α-aluminas upon further calcination at increasing temperatures (49, 50).

Very pure aluminum oxides have been commercially prepared by both metal oxidation and thermal decomposition of aluminum salts. Recently, interest in aluminas forming on a metal surface from the viewpoints of their electronic properties and catalytic activity has increased sharply.

Concluding Remarks

The foregoing discussion of the mechanisms and products of thermal dehydration of aluminum hydroxides and their properties suggests that a wide variety of activated alumina products can be obtained by changing conditions at different stages of preparation. These conditions include the method of preparation of the hydroxide, the dehydration conditions (rate of heating, water vapor pressure, time, and final temperature), and the forming variables. In addition, conditions followed during subsequent operations such as rehydration, aging, and reactivation can also influence final product characteristics. The relationship between structure and catalytic activity of alumina surfaces has also become a subject of considerable interest. Subsequent chapters will show that industrial activated alumina products and applications that are generally available at present cover only a small area of this vast and complex interaction between precursor materials, preparation methods, and activation variables. Enormous possibilities thus exist for the development, production, and uses of many additional activated alumina products of widely differing properties.

Literature Cited

1. Ulrich, F. *Nor. Geol. Tidsskr.* **1925,** *8,* 115–122.
2. Stumpf, H. C.; Russell, A. S.; Newsome, J. W.; Tucker, G. M. *Ind. Eng. Chem.* **1950,** *42,* 1398–1403.
3. Papée, D.; Tertian, R.; Biais, R. *Bull. Soc. Chim. Fr.* **1958,** 1301–1310.
4. Day, M. K. B.; Hill, V. J. *J. Phys. Chem.* **1953,** *57,* 946–950.
5. Ginsberg, H.; Huttig, W.; Strunk-Lichtenberg, G. *Z. Anorg. Allg. Chem.* **1957,** *293,* 33–46, 204–213.
6. Lippens, B. C.; Steggerada, J. J. *Physical and Chemical Aspects of Adsorbents and Catalysts*; Linsen, B. G., Ed.; Academic: New York, 1970; 171–211.
7. Tertian, R.; Papée, D. *J. Chim. Phys. Phys.-Chim. Biol.* **1958,** *55,* 341–353.
8. Brindley, G. W.; Choe, J. O. *Am. Mineral* **1961,** *46,* 771–785.
9. Saalfeld, H. *Neues. Jahrb. Mineral., Abh.* **1960,** *95,* 1–87.
10. Geller, S. *J. Chem. Phys.* **1960,** *33,* 676–677.
11. Saalfeld, H. *Clay Miner. Bull.* **1958,** *3,* 249.
12. Leonard, A. J.; Semaille, P. N.; Fripat, J. J. *Proc. Br. Ceram. Soc.* **1969,** 103.
13. Rooksby, H. P. *Silic. Ind.* **1959,** 335–339.
14. Wefers, K. *Erzmetall* **1962,** *15,* 339–343.
15. Schwiersch, H. *Chem. Erde* **1933,** *8,* 252–315.
16. Deflandre, M. *Bull. Soc. Fr. Mineral.* **1932,** *55,* 140–165.
17. Schwiersch, H. *Chem. Erde* **1933,** *8,* 252–315.
18. Lima-de-Faria, J. *Z. Kristallogr., Kristallgeom., Kristallphys., Kristallchem.* **1963,** *119,* 176–203.
19. Wilson, S. J. *J. Solid State Chem.* **1979,** *30,* 247–255.
20. Lippens, B. C.; DeBoer, J. H. *Acta Crystallogr.* **1964,** *17,* 1312.
21. Sasvari, K.; Zalai, A. *Acta Geol. Acad. Sci. Hung.* **1957,** *4,* 415.
22. Ball, M. C.; Taylor, H. F. W. *Mineral. Mag.* **1961,** *32,* 754.
23. Wilson, S. J.; McConnell, J. D. C. *J. Solid State Chem.* **1980,** *34,* 315–322.
24. Feitknecht, W.; Wittenbach, A.; Buser, W. *React. Solids, Proc. Int. Symp., 4th, 1960* **1961,** 234–239.
25. Rouquerol, J.; Rouquerol, F.; Ganteaume, M. *J. Catal.* **1975,** *36,* 99–110.
26. Rouquerol, J.; Rouquerol, F.; Ganteaume, M. *J. Catal.* **1979,** *57,* 222–230.
27. DeBoer, J. H. *Angew. Chem.* **1958,** *70,* 383–389.
28. Eyraud, C.; Goton, R. *J. Chem. Phys.* **1954,** *51,* 430–433.
29. Eyraud, C.; Goton, R.; Prettre, M. *C. R. Hebd. Seances Acad. Sci.* **1954,** *238,* 1028–1031.
30. Eyraud, C.; Goton, R.; Prettre, M. *C. R. Hebd. Seances Acad. Sci.* **1955,** *240,* 1082–1084.
31. Jackson, G. V.; Jones, P. *Fire Mater.* **1978,** *2* (1), 37–38.
32. Goton, R. Ph.D Thesis, University of Lyon, July 1955.
33a. Yokokawa, T.; Kleppa, O. J. *J. Phys. Chem.* **1964,** *68,* 3246–3249.
33b. Misra, C. Alcoa, unpublished work.
34. Borer, W. J.; Gunthard, H. H. *Helv. Chim. Acta* **1970,** *53,* 119–120.
35. Sabatier, G. *Bull. Soc. Fr. Mineral. Crystallogr.* **1954,** *77,* 1077–1083.

36. Russell, A. S.; Cochran, C. N. *Ind. Eng. Chem.* **1950**, *42*, 1336–1340.
37. Blanchin, L. Ph.D. Thesis, University of Lyon, June 1952.
38. Gregg, S. J.; Sing, K. S. W. *J. Phys. Colloid Chem.* **1951**, *55*, 592–597, 597–604.
39. Lippens, B. C.; de Boer, J. H. *J. Catal.* **1965**, *4*, 319.
40. Kotanigawa, T.; Yamamoto, M.; Utiyama, M.; Hattori, H.; Tanabe, K. *Appl. Catal.* **1981**, *1*, 185–200.
41. DeBoer, J. H.; Lippens; B. C. *J. Catal.* **1964**, *3*, 38.
42. Lippens, B. C. Ph.D. Thesis, Delft University, 1961.
43. Moscou, L.; Lub, S. *Powder Technol.* **1981**, *29*, 45–52.
44. Bedford, R. E.; Berg, M. U.S. Patent 405 072, Sept. 27, 1977.
45. Tischer, R. E. *J. Catal.* **1981**, *72*, 255–265.
46. Riekert, L.; Weber, W. *Chem.-Ing.-Tech.* **1977**, *49* (1), 42.
47. Dignam, M. J.; Fawcett, W. R.; Bohni, H. *J. Electrochem. Soc.* **1956**, *103* (4), 209–214.
48. Scamanis, G. M.; Butler, E. P. *Met. Trans.* **1975**, *6A*, 2055–2063.
49. Funaki, K.; Shimizu, Y. *Kogyo Kagaku Zasshi* **1959**, *62*, 788–793.
50. Fricke, R.; Jockers, K. *Z. Anorg. Chem.* **1950**, *262*, 3.
51. *An Atlas of Alumina*; BA Chemicals Ltd., London, 1969.

6

Industrial Production of Activated Aluminas

As discussed in Chapter 5, the thermal dehydration of aluminum hydroxides results in the formation of porous, high-surface-area aluminas. These products, collectively called "activated aluminas", are widely used in adsorption and catalysis where their large surface area, pore structure, and surface chemistry play significant roles. In all cases, the material is "activated" to the porous structure by thermal dehydration to remove hydrous water. Variations in form, surface area, and, to some extent, pore geometry are achieved by using different starting hydroxides, forming methods, and rates, durations, temperatures, and atmospheres of thermal treatment. Rehydration behavior is an important factor in several commercial production processes. Because activated aluminas are produced in many forms, they can be conveniently classified on the basis of the starting hydroxide.

Activated Bauxites

Activated bauxite is one of the oldest products in this group. Activated bauxite is produced by dehydrating high-alumina-content gibbsite bauxite in the temperature range of 400–800 °C. Appropriate crushing and screening of feed and product produces a granular product in the size range of 5–75 mm. The product contains 10–30% impurity oxides, comprising mainly Fe_2O_3, SiO_2, and TiO_2. The final water content (determined as weight loss on ignition to 1200 °C) is controlled between 4 and 10%. The material has a BET surface area of 75–150 $m^2 g^{-1}$.

Activated bauxites have been used as decolorizing and moisture-removal agents for natural oils and fats and mineral oils. Activated bauxite's use as a Claus catalyst for sulfur removal from sour gas has been mostly superseded by purer activated aluminas.

Activated Aluminas from Bayer Hydroxide

Bayer aluminum hydroxide is the chief source of commercial activated alumina products. Powder forms of activated alumina are produced by heating

Bayer hydroxide in the temperature range of 300–750 °C. The thermal dehydration process is carried out in tray-equipped ovens and rotary and fluidized bed calciners. The products have specific surface areas of 200–350 m² g^{-1} and losses on ignition of 3–12% by weight. Such products are used as decolorizing agents for organic chemicals and as starting materials for the production of aluminum fluoride. Laboratory uses of activated alumina powders include chromatographic and catalytic applications in organic chemistry. For some uses, soda must be removed by washing the activated product with water or dilute acids followed by drying and reactivation.

Granular activated alumina produced from Bayer plant crust (or scale) is one of the oldest commercial forms of this product. This alumina is still widely used. In the Bayer operation for bauxite refining (see Bayer Process in Chapter 3), aluminum hydroxide crystallizes out on the walls of process vessels, particularly in precipitator tanks; these crystals form a hard deposit up to 30 cm in thickness. This crust is periodically removed from vessel walls by mechanical breakers. The large pieces are crushed and screened to the desired grain size in the range of 5–20 mm. The product then is activated by heating to 400–600 °C in a current of hot air or combustion gases. A flow sheet of the process is shown in Figure 6.1. The activated material is a hard, nondusting product. Table 6.I lists some properties of a commercial crust-based activated alumina product (Alcoa F-1). X-ray diffraction analysis of the product shows a pattern of χ- and γ-Al_2O_3 with minor amounts of boehmite. The product has a broad pore distribution up to 5000 Å in diameter,

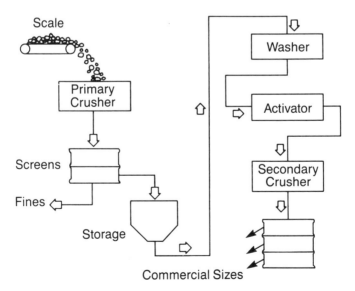

Figure 6.1. Flow sheet for production of activated alumina from Bayer plant crust.

Table 6.I.
Typical Properties of Commercial Activated Aluminas

Properties	F-1	H-151	S-100
Al_2O_3 (%)	92	90	95
Na_2O (%)	0.58	1.6	0.35
SiO_2 (%)	0.12	2.0	0.03
Fe_2O_3 (%)	0.06	0.03	0.05
LOI (%)	7	6	5
Loose bulk density			
g/cm^3	0.83	0.82	0.80
lb/ft^3	52	51	50
Packed bulk density			
g/cm^3	0.85	0.85	0.75
lb/ft^3	55	53	47
Helium (true) density (g/cm^3)	3.25	3.40	3.15
Mercury (particle) density (g/cm^3)	1.42	1.38	1.24
Micro pore volume (pores < 35 Å) (cm^3/g)	0.17	0.23	0.12
Macro pore volume (pores > 35 Å) (cm^3/g)	0.23	0.20	0.37
Total pore volume (cm^3/g)	0.40	0.43	0.49
Total porosity (%)	56.3	59.4	60.6
Pore diameter at 50% total pore volume (Å)	177	35	47
Primary pore size range (Å)	0–100,000	0–400	0–5000
BET surface area (m^2/g)	250	360	260
Pore diameter (Å)			
10^6–10^7 cm^3/g pore volume	0.031	0.003	0.000
10^5–10^6 cm^3/g pore volume	0.045	0.001	0.001
10^4–10^5 cm^3/g pore volume	0.059	0.001	0.004
10^3–10^4 cm^3/g pore volume	0.029	0.004	0.033
10^2–10^3 cm^3/g pore volume	0.045	0.093	0.047
35–10^2 cm^3/g pore volume	0.021	0.099	0.289
2–35 cm^3/g pore volume	0.165	0.232	0.115
Total	0.395	0.433	0.489

although the major part, about 70%, of the surface area is accounted for by the smaller, less than 50-Å-diameter, pores (Figure 5.8).

Due to changes in tank descaling practices resulting from the high cost and handling, safety, and noise problems associated with mechanical descaling, the continued commercial availability of this product is uncertain.

A similar granular product has been produced by compacting Bayer hydroxide by mechanical pressure. The process [developed and commercialized by Martinswerk GmbH in West Germany (1)] utilizes a roll compactor (or granulator). The product from the compactor is broken up and sieved to the required size; oversize and undersize materials are recycled. The screened-size fractions are activated in the temperature range of 400–600 °C in a rotary calciner. Both external firing (indirect) and internal firing (gas or oil) of the calciner have been used for activation. Minor variations of product characteristics can be effected by the firing technique, the air and combustion gas flow, and the residence time of the material in the different temperature zones of the calciner. The activated products are normally transported in sealed steel drums.

In general, the product from the compaction process is dustier and the crushing strength is somewhat lower than those of the Bayer crust material. Other properties such as density and surface area are nearly identical. The compaction process is claimed to be responsible for the formation of a secondary, large-diameter pore structure in addition to the small, less than 30-Å primary pores generated by the release of hydrous water during activation.

The majority of commercially important activated alumina products available today are produced by the rapid dehydration processes. This development followed the identification of rehydration behavior of amorphous ρ-alumina, which is obtained by the rapid removal of water vapor during thermal dehydration of aluminum trihydroxides.

The development of this process provided solutions to two major problems of preparing industrially useful (for adsorbent and catalytic applications) activated alumina products from relatively inexpensive and abundant Bayer aluminum hydroxide. As mentioned under Dehydration Mechanisms in Chapter 5, the thermal dehydration of Bayer hydroxide can result in the formation of 10–25% hydrothermal boehmite. This formation has been shown to be the cause of the decrease in the surface area of the activated product compared to that expected from complete transformation of gibbsite to the transition (χ-Al_2O_3) form. Adsorption and catalytic processes also are generally carried out by using packed beds of preferably spherical particles of the desired size. Hence, a need to prepare the activated alumina in this form exists. The rehydration behavior of amorphous alumina produced by fast dehydration of Bayer aluminum hydroxide provided a method for producing a high-surface-area activated alumina of the desired shape and high mechanical strength to satisfy conditions encountered in industrial processes. The pore structure of these products is also amenable to some manipulation by changing forming and other conditions of manufacture. Another favorable factor is the structure of the final activated product. As discussed under Rehydration in Chapter 5, the products of rehydration and aging of the "flash"-activated hydroxide are mainly pseudo-boehmite and bayerite. These forms convert to η-Al_2O_3 on dehydration. There is evidence (2) that η-Al_2O_3 and γ-Al_2O_3 possess characteristically distinct chemisorption properties. The η phase is generally associated with higher surface areas than that of γ-Al_2O_3 at a higher degree of dehydration (lower LOI); its heat of immersion in water is also higher. The surface acidity has been found to be higher,

6. INDUSTRIAL PRODUCTION OF ACTIVATED ALUMINAS 101

and studies have shown higher catalytic activity of η-Al_2O_3 compared to that of γ-Al_2O_3 in several model catalytic reactions. These factors have helped to establish the rapid dehydration route as the most important process for the production of activated alumina products.

One of the earliest reports of the use of rehydratable alumina for the preparation of formed activated alumina products is contained in the patent disclosures by Pingard (3). Pingard proposed that rapid removal of water vapor formed during dehydration should be carried out either by performing the process under high vacuum or by passing a current of hot air at a temperature of about 500 °C through a bed of the hydroxide. The product is finely ground and the ground material formed into spherical agglomerates. Rehydration with water results in the recrystallization of pseudo-boehmite and bayerite, causing the agglomerates to harden. The last stage of the process consists of reactivation of the hard balls at above 300 °C. The activated spherical product, practically free from boehmite, has a surface area in the range of 300–350 m^2 g^{-1}. The X-ray diffraction of the activated product is very close to that of η-Al_2O_3.

Further developments of this process have been in the direction of increasing the dehydration temperature to above 800 °C and correspondingly reducing the contact time to less than a second. Practically instantaneous dehydration takes place in such cases, and the highly amorphous product obtained has improved rehydration characteristics. A strong activated product with a surface area of between 320 and 380 m^2 g^{-1} is obtained after agglomeration and reactivation.

The patent literature contains several examples of equipment for the fast dehydration step. A typical configuration is shown in Figure 6.2 and is taken from Saussol (4), also of Pechiney. A German process described by Podschus (5) of Bayer AG, similar to the process described, has been used for commercial production of ball-shaped activated alumina products.

Many other process improvements and product variations have been claimed in the patent literature. Tertian and Papée (6) and Miller (7) claim higher surface area and improved sorptive capacity by mixing aluminum trihydroxide with the dehydrated material prior to agglomeration, aging, and reactivation. Use of a finer hydroxide, obtained by grinding or precipitation, as feed to the fast dehydration process has been claimed to produce a more active rehydratable alumina. Osment and Emerson (8) proposed the direct introduction of the hydroxide into the combustion (flame) zone of a fuel-fired flame. The flame is maintained at a temperature in the range 1650–1950 °C. Immediately after passage through the flame, the material is rapidly quenched to a temperature of 300–400 °C. The material is further cooled and ground to about 80% −44-μm size. The ground product is mixed with water and agglomerated. The agglomerates are aged for sufficient time for rehydration and development of strength before final reactivation at 350–400 °C.

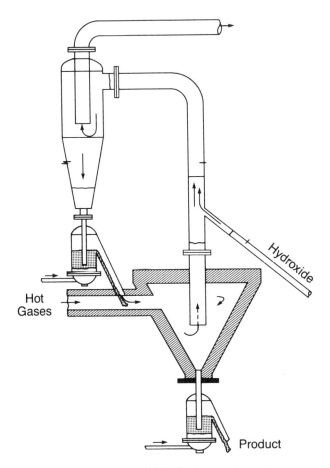

Figure 6.2. Rapid dehydration equipment.

Properties of a commercial activated alumina prepared by the fast dehydration process (Alcoa S-100) are given in Table 6.I.

Activated Alumina from Boehmite and Gelatinous Aluminum Hydroxide

The formation of boehmite as a byproduct of the Ziegler process for linear alcohol production has been discussed under Byproduct of Linear Alcohol Production in Chapter 3. This material is finding increased use for the preparation of activated aluminas, especially for catalytic applications because of the absence of soda normally associated with products derived from Bayer hydroxide. The fine-particle boehmite obtained in this process is extruded to various shapes with or without the use of a binder material. The material is claimed to serve as its own binder when peptized with glacial acetic acid and mulled; this procedure imparts excellent green strength to the extrudate. The extrudate is cut to the required size, dried, and activated at 500-600 °C. The surface area of the activated product is 185–250 $m^2\ g^{-1}$, and the X-ray diffraction pattern corresponds to that of γ-Al_2O_3.

Alumina gels have also been used for the manufacture of activated aluminas. These gels are produced by neutralization of aluminum sulfate or ammonium alum by NH_3 or from sodium aluminate by neutralization with acids, CO_2, $NaHCO_3$, and $Al_2(SO_4)_3$. The gelatinous precipitate is filtered and the filter cake thoroughly washed. The cake is either dried and pulverized or directly extruded and cut to the desired shape and size and then dried.

Spray-drying as well is used to provide dried, spherical particles of about 50-μm size. The dried powder may be agglomerated to yield a spherical product or pressed into pellets. Dehydration of the gel in a hot oil bath also has been used to produce spherical particles. The spherical form in this case results from surface tension effects, and spheres so formed are hardened by partial drying.

The gels in various shapes are activated at 400–600 °C in the same manner as before to an LOI value of about 6%. The activated products usually have an X-ray diffraction pattern of broad, diffuse bands of γ/η-Al_2O_3. Impurities include Na_2O, SO_3, Cl, etc.; the impurities present depend on the neutralization process. Although gels of varying texture can be prepared, the usual industrial adsorbent products have very small pores of less than 40-Å diameter and surface areas in the range of 300–400 $m^2\ g^{-1}$.

Several reactor arrangements have been used to carry out the neutralization reaction and subsequent treatment of the gelatinous product to improve filterability and washability. Poezd et al. (9) have described a continuous-flow reactor that is claimed to allow flexibility and improved control over the hydroxide precipitation reaction. The precipitation apparatus (Figure 6.3) consists of a cylindrical vessel provided with a mechanical stirring device. A second, smaller vessel is located inside at the bottom. This vessel is open at the top and is provided with inlet pipes for the reactants. Solutions of aluminum salts and alkali are delivered continuously into the inner vessel where it forms a concentrated suspension of the hydroxide. This suspension flows out to the bigger vessel where it is diluted with high-purity water. Residence time and temperature of the suspension are controlled to obtain the desired properties of the precipitate.

A two-stage reactor system has been described by Bell et al. (10) for the continuous neutralization of alkali aluminate with mineral acid. The method is claimed to be capable of yielding an aluminum hydroxide of substantially pseudo-boehmite structure. This product, after drying, has a Na_2O content of less than 0.03 wt % and a surface area of 200–300 $m^2\ g^{-1}$.

Partial dehydration of the dried gel followed by agglomeration is the most widely used method for forming spherical products from this material. In the procedure described by Beiding et al. (11), the pseudo-boehmite hydroxide is partially dehydrated to an LOI of 22–30%, ground to at least 85% −325 mesh (44 μm), and shaped into spheres by conventional agglomerat-

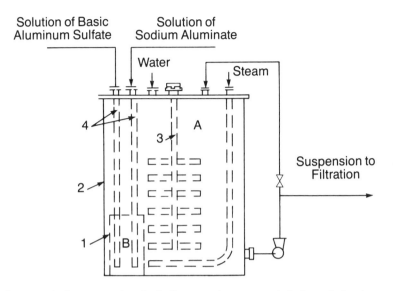

Figure 6.3. Apparatus for single-flow continuous precipitation of aluminum hydroxide: 1, precipitation vessel; 2, dilution vessel; 3, mechanical stirrer; and 4, tubes for delivery of solutions.

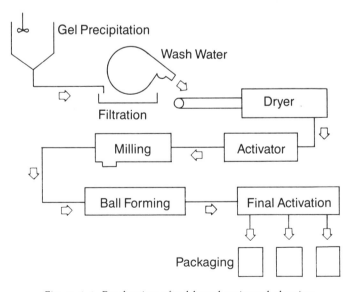

Figure 6.4. Production of gel-based activated alumina.

ing devices while sufficient water is added to obtain agglomerates with a total water content (as given by LOI) of 52–65%. These shaped spheres, without any aging treatment, are reactivated at 350–650 °C for a time necessary to produce desired strength. The resultant spheres have a high porosity, at least 20% of which consists of pores in the 120–800-Å size range.

Table 6.I reports data on a commercial (Alcoa H-151) gel-based product. The manufacture of this spherical product involves a ball-forming (pan agglomerator) operation applied to the milled, partially dehydrated alumina powder followed by a reactivation step. A flow sheet of the process is shown in Figure 6.4. Figure 6.5 shows the appearance of granular and ball-shaped activated alumina products.

Activated Alumina from Bayerite

Activated alumina produced from bayerite is preferred in some catalytic applications. Bayerite is considered to be relatively easier to form (extrusion,

Figure 6.5. Ball-shaped and granular activated alumina products.

pelletizing, tableting, etc.) to the desired geometry. Methods of preparation of bayerite have been discussed under Bayerite in Chapter 3. Bayerite is normally finely ground and then extruded or tableted to the desired shape with a binder. Thermal dehydration at temperatures of 300–600°C yields a product consisting mainly of η-Al_2O_3 with small amounts of boehmite and γ-Al_2O_3.

Literature Cited

1. *Compalox Activated Alumina*; Martinswerk GmbH, 1976.
2. Maciver, D. S.; Tobin, H. H.; Barth, R. T. *J. Catal.* **1963**, *2*, 485.
3. Pingard, L. G. U.S. Patent 2 881 051 to Pechiney, France, 1959.
4. Saussol, F. U.S. Patent 2 915 365 to Pechiney, France, 1959.
5. Podschus, E. German Auslegeschrift 2 059 946 to Bayer AG, 1974.
6. Tertian, R.; Papée, D. U.S. Patent 2 876 068 to Pechiney, France, 1959.
7. Miller, F. E. German Offenlegungschrift 2 020 050, 1972; Br. Patent 1 295 133 to Aluminum Company of America, 1972.
8. Osment, H. E.; Emerson, R. B. U.S. Patent 3 222 129 to Kaiser, U.S.A., 1965.
9. Poezd, N. P.; Kolesnikov, I. M.; Radchenko, E. D. *Zh. Prikl. Khim. (Leningrad)* **1982**, *55 (10)*, 2174–2181.
10. Bell, N.; Price, J. W.; Rigge, J. U.S. Patent 3 630 670 to Kaiser, U.S.A., 1971.
11. Beiding, W. A.; Emerson, R. B.; Williams, R. L. U.S. Patent 3 714 313 to Kaiser, U.S.A., 1973.

7

Activated Alumina: Adsorbent Applications

A major use of activated alumina is in the field of adsorption where its high surface area, pore structure, physical characteristics, and chemical inertness are factors favoring its applications. Important technical applications include gas and liquid drying, water purification, selective adsorption in the petroleum industry, and chromatographic separation processes.

Activated aluminas have been used as adsorbents for many years. In the United States, Alcoa introduced activated aluminas commercially in 1932 (1). These products were used mostly as desiccants in the 1940s and 1950s. The chemical industry is a large consumer of activated aluminas mainly for gas and liquid drying applications, although the earliest known use of activated alumina was in the chromatographic purification of liver extracts (2). Alumina was the common column packing material used in laboratory-scale chromatographic separations from its beginnings in the 1930s. Uses of adsorbent aluminas were expanded in the following years to the isotopic separation of actinide series compounds as well as some organic compounds. This expansion was followed by applications in the areas of separation of halogenated compounds from aqueous streams and organic acids from hydrocarbons (3). The chronological development of uses of activated aluminas is summarized in Table 7.I, which is taken from a recent review paper by Goodboy and Fleming (4).

In recent years, nondesiccant uses of adsorbent aluminas have grown enormously; desiccant uses now account for less than half of the total yearly adsorbent alumina consumption in the United States. This shift is largely due to the growth in adsorptive separation processes, particularly for biological materials, and availability of aluminas designed for greater specificity in the separation of complex organic mixtures. A better understanding of surface chemistry and pore structure and development of methods to control these properties have contributed to this shift from traditional desiccant applications to more complex adsorptive separation processes.

0065-7719/86/0184/0107$07.25/1
© 1986 American Chemical Society

Table 7.I.
Development of Commercial Alumina Adsorption Applications

Year	Molecular Weight	Gas Phase Impurity	Gas Phase Product	Liquid Phase Impurity	Liquid Phase Product
Pre-1940	low	water	—	—	—
1960s	low	water	—	water	—
		isotopes	isotopes	halogens	—
				color bodies	—
	high	—	—	organometallics	—
1980s	low	water	—	water	—
		isotopes	—	halocarbons	—
		SO_2	—	halogens	—
		H_2S	flavors	phosphates	—
		acid gases	fragrances	inorganic acids	—
		amines	xylenes	ketones	—
		—	aromatics	carboxylic acids	—
				mercaptans	—
				alcohols	—
				sulfides	—
				metals	metals
				color bodies	—
	high			organometallics	—
				pharmaceuticals	pharmaceuticals
				$>C_5$ paraffins	sugars
				—	vitamins

Gas Drying

Drying of gases is one of the most important industrial applications of activated alumina. One reason for this is the strong affinity of activated alumina for water.

The mechanism of binding of water on the activated alumina surface still remains unclear. Present thinking favors a mechanism of dissociation of water to H^+ and OH^- ions that become attached to surface sites consisting of an oxide ion on the outermost surface layer and an incompletely coordinated aluminum ion in the next-lower layer. This exposed cation is located in a "hole" that is electron-deficient and thus acts as a Lewis acid site. The rehydroxylation of the dehydroxylated alumina surface on exposure to water vapor is accompanied by considerable heat evolution. This heat amounts to as much as 100 kcal/mol of H_2O at the lowest coverages (5a); a strong chemical interaction between surface and water molecules must be occurring.

Desiccant Properties. CAPACITY. The static water adsorption capacity, also known as equilibrium capacity, is the maximum amount of water that is taken up by the desiccant. This capacity is expressed as a fraction or percentage of the original weight of the desiccant. Equilibrium capacity is obviously dependent on the partial pressure of water vapor (usually expressed as relative humidity) in the gas and the temperature. The data are usually presented in the form of an adsorption isotherm, showing weight percent of water adsorbed versus percent of relative humidity. The equilibrium water adsorption capacities of some typical commercial activated aluminas (Alcoa F-1, H-151, and S-100) in contact with air (25 °C) at different relative humidity conditions are shown in Figure 7.1.

Another useful property of a desiccant is the value of the final residual water content of the gas attainable by the desiccant. As can be seen from Figure 7.2 (5b), activated alumina can dry the gas to a lower water content (commonly expressed as the dew point) than any other commercially available desiccant.

HEAT OF WETTING AND ADSORPTION. The adsorption of water on activated alumina is strongly exothermic. Experimental determination of this heat generation is normally reported as heat of wetting by water. This property has been reported in several different ways, such as "average", "integral", and "differential" heats of wetting, resulting in some confusion.

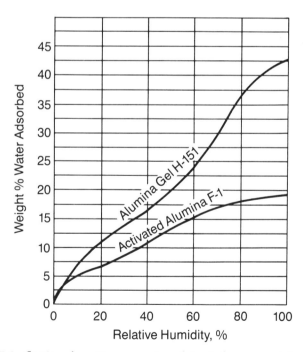

Figure 7.1. Static adsorption capacity of typical commercial activated aluminas at 25 °C.

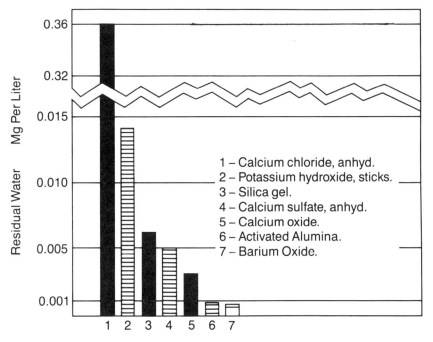

Figure 7.2. Comparative efficiency of moisture removal of various desiccants. Reproduced with permission from Reference 5b. Copyright 1944 National Bureau of Standards.

Stowe (6) determined the heats of wetting by water of several commercial and experimental activated aluminas. The integral heat of wetting, which is the total heat produced when immersing a unit mass of adsorbent in water, was determined to range between 14 and 24 cal/g of fresh adsorbent. The differential heat of wetting by water was determined by measuring the heat of wetting of the adsorbent, which already contained various known quantities of uniformly adsorbed water. The differential heat of wetting decreases sharply as the loading of water increases; the value calculated by Stowe was 170 ± 21 cal/g of water for wetting by a monolayer of water derived from liquid water.

A later work by Venable et al. (7) reports results for integral and differential heats of wetting of several samples of alumina determined from calorimetric measurements of the heats generated by immersion of alumina powders preequilibrated with various amounts of water vapor. At coverages below 0.2 monolayer, the differential heat curves were approximately linear and decreased rapidly from the extrapolated zero coverage value of 21 kcal/mol. In the region of 0.2–0.3 monolayer, a change in slope occurs where the differential heats still decrease but much less rapidly than before. A second transition region occurs in the vicinity of 0.7–0.8 monolayer; this transition is followed by a third, approximately linear portion extending almost to the completion of the second layer. After completion of the second layer, the differential heat relative to the liquid state is practically zero.

Cornelius et al. (5) reported values of 10 kcal/mol of water adsorbed at 3% by weight water adsorption level to over 105 kcal/mol at the very low loading of about 0.1% adsorption level. According to Sanlaville (8), the differential heat adsorption is 6–7 kcal/mol for 0.30 mg of H_2O/g of Al_2O_3 (first layer), 0.9 kcal/mol for 130 mg of H_2O/g of Al_2O_3 (second layer), and 0.25 kcal/mol for more than 150 mg of water adsorbed per g of alumina.

The average heat of wetting (H) is normally reported in commercial literature. This heat is related to the differential and integral heats by the relationships

$$\Delta H(\text{differential}) = d(\Delta H)/dw$$

or

$$\Delta H(\text{integral}) = \int_0^w \Delta H \, dw$$

and

$$\Delta H_{av} = \int_0^w \Delta H \, dw/w$$

where w is the loading of water on the adsorbent.

Average heat of wetting values reported for commercial activated aluminas are in the range of 10–13 kcal/mol of water adsorbed. For adsorption of moisture from a gas, the heat of condensation of the water vapor must be added to the heat of wetting to arrive at the total heat generation value.

ADSORBENT FORM AND CHARACTERISTICS. Both granular and spherical forms of activated alumina are used in desiccant beds. Granules and ball forms from 2 to 20 mm in size are available commercially. Properties of some commercial grades are given in Table 6.I.

Dustiness (abrasion resistance) and crushing strength are practical considerations because they affect the usable life of the desiccant bed. Fines, produced by breakdown, attrition, decrepitation, and flaking cause clogging of the activated alumina bed, which increases the pressure drop. Alumina fines entrained by the dried gas can also cause problems in subsequent process steps.

Surface area and pore characteristics vary depending on the manufacturing process. The granular, lower surface area products are relatively cheaper, and their application is largely based on economy (low cost per pound), low water load, and low stream pressure. Spherical products, produced by the gel or fast dehydration processes, have larger surface areas, have a narrower pore structure, and are relatively more expensive. These products have higher adsorption capacity as shown in the examples of Figure 7.1. These products are usually specified for high-pressure, high-moisture removal duties.

Pressure drop across the desiccant bed is another important consideration. Spherical products have a much lower pressure drop than the irregular granular material and hence are preferred when a low pressure drop is a process requirement. Pressure drop through a packed bed of activated alumina can be estimated by using chemical engineering formulas for flow-through packed beds. The Brownell (9) method is frequently used. In general, the pressure drop per unit height of bed (ΔP) can be satisfactorily represented by

$$\Delta P = aV + bV^2$$

where V is the nominal gas velocity (for an empty column) and a and b are parameters dependent on particle diameter, bed void, gas viscosity, and density. The linear velocity is normally kept below about 1 m s^{-1} to avoid bed expansion and shifting due to the onset of fluidization.

Gases. A list of gases that can be dried by alumina desiccants is given in the box. Relatively few gases will chemically attack activated alumina. This is the reason for its broad application in gas drying operations. In many applications, the alumina desiccant can dry the gas to a lower dew point than any other commercially available desiccant.

Refrigerants must be dried to moisture contents of below 30 ppm to prevent freeze-ups and corrosion in refrigerant lines. Alumina not only dries the refrigerant below 10-ppm water content but also adsorbs acids formed by reaction between refrigerant and water.

Gases Dried by Activated Alumina Desiccant

Acetylene	Cracked gas	Hydrogen	Oxygen
Air	Ethane	Hydrogen chloride	Propane
Ammonia	Ethylene	Hydrogen sulfide	Propylene
Argon	Freon	Methane	Sulfur dioxide
Carbon dioxide	Furnace gas	Natural gas	
Chlorine	Helium	Nitrogen	

7. ADSORBENT APPLICATIONS OF ACTIVATED ALUMINA

Industrial Practice. Most industrial gas-drying operations take place under dynamic conditions. In a typical gas-drying system, moisture is adsorbed at high efficiency when a freshly reactivated desiccant is used. How closely the moisture removal efficiency approaches 100% will depend on such factors as inlet gas temperature, relative humidity or degree of saturation of the gas, desiccant bed depth, desiccant granule size, gas flow rate, and total pressure in the system. As the adsorbent gets progressively more loaded with moisture, removal efficiency decreases until it reaches zero when equilibrium is attained.

In commercial practice, the desiccant is not allowed to come to equilibrium. The desiccant is reactivated when the efficiency has dropped to a point where the maximum allowable water content is passing through the bed. Drying is generally performed by passing the moist gas through a column (or tower) packed with the adsorbent. Most of the moisture in the gas is adsorbed at the inlet region of the column. A layer is formed in which the adsorption equilibrium tends to become established between the adsorbent and the inlet gas.

In the case of alumina adsorbents the maximum amount of water that can be held by the adsorbent layer is given by the equilibrium curve for the relative humidity in question. Most of the water content of the resulting amount of gas can only be retained in the next portion of the column. Thus, the adsorption zone progresses along the column during the operation. Drying efficiency decreases sharply when the adsorption front reaches the end of the column. The drying operation must be stopped before reaching this point to ensure that the required maximum dew point is not exceeded. Operation of two columns connected in series is a practical way of avoiding this problem. The first column can then be loaded close to its equilibrium capacity, and at the same time the maintenance of the specified dew point of the dried gas is ensured. Usually, three towers are employed. At any time two of these towers in series are drying while the third one is being regenerated. This method enables loading the towers to their full adsorptive capacity.

Adsorption of water on activated alumina is strongly exothermic, releasing between 11 and 13 kcal of heat/mol of water adsorbed. If no means are available to take away this heat, the temperature of the desiccant bed will increase. This increase is undesirable because higher temperature reduces the relative humidity and hence the adsorption capacity of the bed. This effect is particularly bothersome in units drying gas at low pressure because gases do not have sufficient heat capacity to carry away the heat of adsorption. Cooling coils have been installed in desiccant beds to remove heat. Gases under high pressure and liquids have sufficient heat capacity to carry away the heat of adsorption. Drying then occurs under practically isothermal conditions.

The following sections present information relating to the practical application of activated alumina for gas drying. A typical industrial gas-drying installation for natural gas using activated alumina is shown in Figure 7.3. A flow diagram of the process is presented in Figure 7.4.

Split-Bed Design.   A combined split bed of the more active spherical product with less active granular material has been used as a cost-effective design for drying installations. The high-capacity desiccant is placed at the inlet end of the bed, where the relative saturation is highest, to take advantage of its high sorptive capacity at this condition. Locating the less active product at the exit end of the bed, where relative saturation is low, utilizes its ability to

Figure 7.3. Natural gas drying plant.

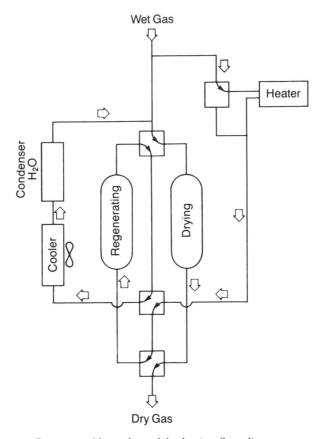

Figure 7.4. Natural gas dehydration flow diagram.

achieve a low dew point. The split-bed design using the two aluminas can provide improved dew point performance at a lower total desiccant cost.

Breakthrough Capacity. In gas-drying practice, the required maximum dew point of the effluent stream is usually specified. A dew point lower than that specified is achieved for some time after starting the drying cycle. Then the leading edge of the adsorption front moves up and the effluent dew point increases. When the dew point reaches the specified limit, the system is said to have attained the breakthrough point. The total amount of water then fixed in the bed divided by the weight of the desiccant represents the breakthrough capacity. The breakthrough capacity for a given dew point, in addition to being dependent on intrinsic properties of the alumina such as surface area, pore size, and pore volume, depends on operating factors such as gas relative humidity, temperature, pressure, flow rate, and contact time. This capacity is related also to loss of performance of the adsorbent due to regeneration and aging.

Regeneration. Regeneration removes sorbed material to bring the alumina back to its activated state. This removal is ideally brought about by passing a stream of hot dry gas through the desiccant bed. At ordinary temperatures, most of the adsorbed water in alumina can be removed either in this manner or by degassing under vacuum. However, for removal of the last traces of water strongly held on the alumina surface, the temperature must be raised and at the same time a low water vapor pressure must be maintained. Figure 7.5 shows the amount of residual water held at equilibrium by an adsorbent alumina for varying conditions of regeneration. Practically all the adsorbed water can be removed by a current of gas, even if moist, provided the gas is at a sufficiently high temperature.

Heat requirements for regeneration include thermal capacity of the adsorbent bed and the heat of desorption (including the latent heat of evaporation of the adsorbed water). This heat must be provided by the gas used for regeneration or some other heat input source. The gas flow must be sufficient to carry away the desorbed water under low relative humidity conditions in order to bring the residual water in the adsorbent to a sufficiently

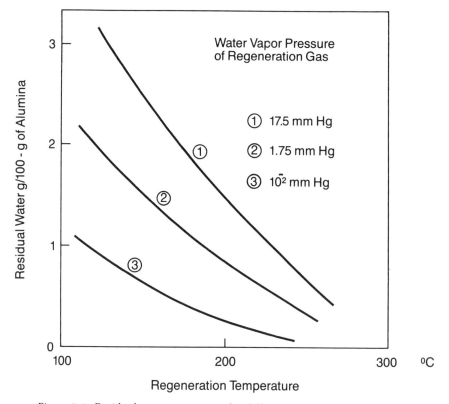

Figure 7.5. Residual water content under different regeneration conditions.

low level. The gas used for regeneration can be either air or the gas being dried. At the end of the regeneration operation, the alumina has to be cooled to normal temperature without introducing significant amounts of water.

An idealized drying–regeneration cycle is shown in Figure 7.6. The cycle is represented by (1) adsorption of moisture at operating temperature t_1 (line AB), (2) desorption (regeneration) by raising the temperature to t_2 at constant water vapor pressure (line BC), and (3) cooling of the bed at constant water content (line CA).

The amount of water left in the bed at the end of the regeneration cycle governs the drying efficiency in the next cycle. When a very high efficiency is required, a dry gas and a regeneration temperature over 150 °C are normally employed. Where only a limited amount of regenerating gas is available, it must be at a sufficiently high temperature to provide the required quantity of heat to desorb the fixed water. If a sufficiently large amount of dry gas is available, a lower temperature can be used. Electrical heaters or steam coils imbedded in the adsorbent can also be used for regeneration. However, unless carefully controlled, this procedure could lead to local overheating and degradation of adsorbent.

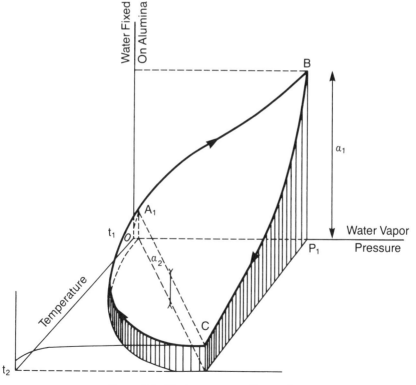

Figure 7.6. Drying–regeneration cycle.

Regeneration is usually carried out countercurrent to the gas flow during the drying cycle. This procedure ensures effective regeneration while minimizing aging effects caused by contact of alumina with hot wet gases. Aging effects are further minimized by using very short regeneration cycles and purging with dry gas under reduced pressure.

Fouling and Aging. A typical example of desiccant fouling is carbonaceous deposits that form on the alumina when drying in natural gas and unsaturated hydrocarbons. Other causes of fouling include deposition of droplets of oil and tar and, due to degradation, polymerization or oxidation of unstable compounds present in the gas. These deposits effectively plug the pores, making them unavailable for moisture adsorption. Over a period of time, these deposits, if not driven off during routine reactivation, eventually lower the adsorptive capacity of the desiccant to a point where it can no longer be used. However, this effect has been shown to be less severe in the case of activated alumina compared to other adsorbents such as molecular sieves. These carbon deposits can be burned off by high-temperature (500–600 °C) regeneration of alumina under carefully controlled (to avoid excessive temperatures) conditions. Data presented in Figure 7.7 are an example of the results obtained from high-temperature regeneration tests.

The aging of the adsorbent shows up as a decrease of its drying capacity after several operating cycles. Aging is directly related to the number of re-

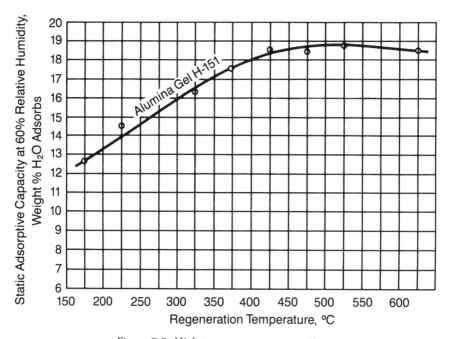

Figure 7.7. High-temperature regeneration.

Table 7.II.
Drying Gases from Steam Cracking of Naphtha

Number of Regenerations	Effluent Water Content (ppm of vol)	Length of Cycle (h)	Breakthrough Capacity (%)
0	0.5	94	20
8	0.5	75	16
102	0.5	47	10
148	0.5	42	9

NOTE: The quality used was activated alumina balls of 2-5 mm. Three driers were used: two operating in series and one operating under regeneration. The drying conditions were as follows: pressure, 35 atm (500 psi); temperature, 15 °C (59 °F); water content, 500 ppm by volume; contact time, 30 s; and required efficiency, less than 3 ppm of water vapor by volume. The regeneration conditions were countercurrent, with light dry gas, free of olefinic hydrocarbons, amounting to 5-10% of the weight of the gas to be dried. The maximum inlet temperature was 240 °C (474 °F).

generations and is the result of irreversible changes in the adsorbent structure. In the case of activated alumina, aging is attributed mostly to hydrothermal phase changes under regeneration conditions. Under usual working conditions, a 30% drop in the starting adsorption capacity is observed in 2-4 years of operation. An initial rapid decrease followed by a slower decline is commonly observed. This factor must be considered when estimating the design capacity for the desiccant in any particular application, particularly where long-term adsorptive capacity is an important consideration.

The service life of activated alumina desiccant has been reported to be as high as 10 years. Replacement is necessary when the dew point or the drying cycle time cannot be maintained or when the pressure drops to an unacceptable level. Data from a typical industrial gas-drying application using activated alumina are presented in Table 7.II.

Liquid Drying

Liquids that can be dried with activated aluminas include aromatic hydrocarbons, heavier alkanes, gasoline, kerosene, cyclohexane, propylene, butylenes, and many halogenated hydrocarbons. The liquid to be dried by alumina must satisfy some general criteria:

1. The liquid must not react or polymerize in contact with the adsorbent. An example of a liquid that cannot be dried by alumina is acetone. Acetone is catalytically oxidized to mesityl oxide and mesitylene in the presence of alumina at elevated temperatures.

2. Compounds that are adsorbed to the same extent as water should not be present in appreciable quantities in the liquid to be dried. In some cases, the affinity of a liquid for the alumina desiccant, as evidenced by the heat of wetting, is as high as that of water. Highly polar compounds (e.g., methyl and ethyl alcohols) are examples of such liquids.

3. The liquid must be free of components that will deposit on the alumina surface and will not be driven off during reactivation. For example, sulfur quickly clogs the pores and reduces adsorption capacity. Sometimes, high-boiling components of a liquid are adsorbed by alumina and are not driven off at normal reactivation temperatures of 150–200 °C. Occasional high-temperature regeneration is useful in these cases.

Regeneration schemes for liquid dehydration units are varied and depend on the liquid being dried. In some applications the liquid being dried is vaporized, heated, and passed through the desiccant bed to desorb water. Gases can also be used for regeneration. Often the gas used, for example, nitrogen, is inert and entirely different from the liquid being dried.

Examples of drying of organic liquids with activated alumina are reported by Derr and Willmore (10). Smith and Leva (11) have discussed the use of granular desiccants for drying liquids by adsorption.

Water Purification

A rapidly evolving application of activated alumina is in the area of water purification. Several important contaminants have been successfully removed from water in both pilot and large-scale treatment plants. Interest in such processes has grown following the passage of clean water legislation and more stringent conservation requirements.

Fluoride. Activated alumina has been used to reduce fluoride concentration in drinking water to meet regulations for control of teeth fluoridolysis. Several treatment plants are in operation in the United States and elsewhere.

High fluoride-removal efficiency is reached at a pH of 5.5 (12). Bed capacities vary from 4.5 to 7.0 kL min^{-1} (m^3 of alumina bed)$^{-1}$. Regeneration is carried out with 1% NaOH solution. The cost of operation, including chemicals, electricity, bed replacement, maintenance, and labor, is estimated to be between 3 and 10¢/m^3 of treated water.

Industrial effluents also have been successfully treated with activated alumina to lower the fluoride content to acceptable levels.

Color and Odor. Activated alumina is effective and cost competitive (e.g., with activated carbon) for removal of color and odor from industrial effluent water. Water from dye works and paper plants has been successfully and economically treated by this method. Regeneration is usually carried out by calcining the spent alumina at 500–600 °C to burn off adsorbed contaminants.

Phosphate. Although investigations on the use of activated alumina to remove phosphate from water have been published (13), no reports of large-scale operations exist. Reports have also been published on the removal of arsenic from water by adsorption on activated alumina (14).

Hydrocarbon Recovery and Selective Adsorption Applications

The pore size distributions and surface chemistry of activated aluminas are conducive to the adsorption and retention of molecules other than water. The high affinity of activated alumina for water causes other molecules (except low molecular weight alcohols) to be displaced by water on a time-dependent basis. If the time for such displacement is long, the amount of hydrocarbons or other adsorbates retained by the alumina may be quite large compared to the amount of water in a dynamic application with short contact time. Thus, the short-cycle dehydration process can be used to recover heavy hydrocarbons from streams of lighter hydrocarbons by adsorption on alumina, followed by thermal desorption. A typical hydrocarbon recovery system is shown in Figure 7.8. The basic steps in the process are

- short-cycle drying,
- short-cycle regeneration,
- stripping with hot gas,
- condensing and separating H_2O and hydrocarbons, and
- recycling stripping gas through the heater and bed.

The amount of hydrocarbons recovered in a short-cycle plant depends on stream composition, efficiency of adsorption and desorption of hydrocarbons from the adsorbent, and efficiency of the condenser system. Case histories have shown that recoveries of approximately 70% of the stream hydrocarbons that are C_5 and higher can be achieved by using activated alumina adsorbents. This figure assumes optimum cycle time and that the regeneration procedure does not promote severe coking. A gel-based spherical alumina is preferred in this application because of its high pore volume and low pressure drop since high-volume gas flows are common in a short-cycle plant.

Activated alumina can be used to separate one or more components from a gas or liquid stream. In most cases, a regeneration scheme can be devised that permits cyclic use of the alumina, but in other instances, separation on a one-shot basis is economical. Examples of such applications include removal of catalyst from polyethylene and in hydrogen peroxide production.

Although no absolute rules govern the separation of components by activated alumina, the more polar molecules are usually removed selectively from the bulk stream. Polar molecules, such as water and alcohols, are highly adsorbed by activated alumina, while nonpolar gases, such as nitro-

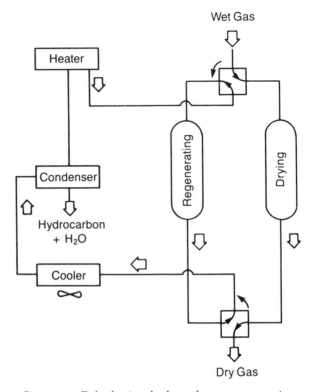

Figure 7.8. Dehydration–hydrocarbon recovery unit.

gen and argon, are only very weakly adsorbed. Activated aluminas can remove small quantities of organic fluorides from gasoline produced by the hydrogen fluoride alkylation process, chlorides from hydrogen, and compressor oil from compressed gases.

The most direct way to determine relative affinities is to conduct bench-scale tests to establish which components are sufficiently adsorbed, retained, and released for economical separation.

Maintenance of Power System Oils

Oil is employed in transformers and oil circuit breakers as a coolant and insulator and in other power system applications as a lubricant. Contact of the warm oil with air and moisture causes the formation of organic acids and sludge. Unless this degradation process is inhibited or the contaminants are removed, the cooling, insulating, and lubricating value of the oil is greatly impaired; equipment damage results in extreme cases.

Granular-activated alumina has been used for the continuous maintenance of power system oils to ensure sustained good performance (15).

Transformers and other oil-using components are equipped with a continuous treater containing a bed of activated alumina through which the oils are passed. Oil flow through the bed is normally accomplished by a thermosiphon system utilizing the temperature difference between equipment and treater vessels. A separate breather, also using activated alumina, is used to prevent moisture entry.

The activated alumina is replaced every 1–4 years when its performance falls below required levels. The treatment process removes moisture and any oxidation products from the oil that are adsorbed by the activated alumina. Reactivation of the alumina is possible by burning off adsorbed sludge and adhering oil at 500–600 °C. However, reactivation is often not economically justified, and the spent alumina is discarded.

Chromatography

Use of alumina in analytical chromatographic separations began around 1930 and is represented by the work of Karrer et al. (*16*). A patent issued to Folkers and Shovel (*2*) as early as 1901 described a process for the purification of liver extract using alumina. Aluminas and silicas were the common chromatographic materials in the 1940s and 1950s. Analytical chromatography using conventional packed columns of alumina for gas and liquid separations was the main area of activity during this period. The years following saw considerable development in chromatographic techniques, but most of these techniques used polymeric materials as the separation medium.

The past 10 years have seen a resurgence of interest in alumina as a versatile material for all types of chromatographic separation processes and products. This resurgence has closely followed the development of chromatography as a commercial separation unit operation. Tables 7.III and 7.IV are collections of representative chemical and biological chromatographic separations that use aluminas as the separation medium. The tables show the diverse nature of products and processes used.

In practice, almost all organic compounds, except perhaps saturated aliphatic hydrocarbons, are adsorbed to some degree on alumina. This adsorption behavior is related to the chemical nature of the alumina surface. Activated aluminas can be considered as possessing both Lewis and Brönsted acidic and basic sites of various strengths and concentrations. Acidity is contributed by coordinately unsaturated Al^{3+} ions, protonated hydroxyls, and some acidic hydroxyls. Basicity is a result of O^{2-} anion vacancies and basic hydroxyls. A summary of adsorption properties of chromatographic alumina is given in Table 7.V.

Commercial chromatographic aluminas are available in neutral, acidic, and basic grades. Neutral alumina has, in general, the widest range of applications and is used for the separation of steroids, alkaloids, hydrocarbons, esters, aldehydes, alcohols, and weak organic acids and bases. This alumina

Table 7.III.
Example of Chemical Chromatographic Separations Using Aluminas

Material	Technique	Reference
Aromatic hydrocarbons	AC[a]	20–23
Halogens and organic halides	AC	24
Xylenes and nitrobenzene	HPLC	25
Benzene and naphthalenes	HPLC	26
Aldehydes	HPLC	27
Polychlorinated biphenyls and dioxins	HPLC	28
Nitrogen heterocycles and coal tar	LC	29

[a]AC is adsorption chromatography.

Table 7.IV.
Example of Biological Chromatographic Separations Using Aluminas

Material	Technique	Reference
Steroids	HPLC	30
Streptomycins	LC	31
Alkaloids (atropine, etc.)	TLC; HPLC	32; 33
Estrogens	LC	34
Peptides	HPLC	35
Testosterone and epitestosterones	TLC	36
Phospholipids	HPLC	37

is usually used with nonaqueous solutions. Basic aluminas, which usually contain some alkali, exhibit strong cation-exchange properties in aqueous media and adsorb basic amino acids, amines, and other basic substances. These aluminas are capable of adsorbing aromatics and similar unsaturated compounds from organic solvents. Acidic alumina is an acid-washed preparation that acts as an anion exchanger and may be applied to the separation of both inorganic anions and acidic organic molecules such as acidic amino acids, aromatic acids, and carboxylic acids.

The activity of chromatographic alumina has been traditionally reported in the Brockman and Schodder (17) scale, which is derived from a simple empirical test for adsorption activity. The highest activity is given by grade I, which probably represents a high level of dehydroxylation. Higher grade numbers represent falling adsorptive activity as a result of increased adsorption of water.

Large-scale, commercial chromatographic separation using alumina has become an established unit operation, particularly in pharmaceutical and biological processing. Some of the large-volume commercial separations include steroids, proteins, and vitamins. The flavor and fragrance industry uses large alumina-packed chromatographic columns for separation

Table 7.V.
Summary of Adsorption Properties of Chromatographic Alumina

Solute	Extent of Surface Coverage	State of Adsorbed Solute	Site Adsorption	Probable Nature of Bond
Nonpolar, nonionic (aromatic hydrocarbons)	very small (each 0.1%)	small isolated clusters of molecules; end-on orientation	Al^{3+}	π-electron complex with surface aluminum atoms
Polar, nonionic				
Proton acceptor	small	monodisperse; end-on or flatwise orientation	$>$Al–OH	hydrogen bonding with surface –OH groups
Proton donor	complete	monodisperse; end-on or flatwise orientation	$>$Al–O$^-$	hydrogen bonding with surface oxygen atoms
Ionic				
Cationic	not known	cationic micelles[a]	Al–O–H$^+$	ion exchange with surface protons
Anionic	complete	monodisperse anions and anionic micelles	Al$^+$X$^-$	ion exchange with surface anions of acidified alumina and some covalent bonding
Chelating	complete[a]	chelate complex	$>$Al$^+$ . . .[a]	lake formation with aluminum ions

SOURCE: Reproduced with permission from Reference 3. Copyright 1965 Reinhold.
[a]This is tentative.

operations. The particle size of alumina used in chromatography varies from 50-µm-diameter (powder) to 0.2-mm-diameter smooth spheres. A very high level of chemical purity of the alumina is normally required in most chromatographic applications.

Recent Developments in Adsorbent Aluminas

In the past, aluminas have generally been considered as adsorbents with limited adsorption capability and of low specificity. This image has changed in recent times with the development of products that are better characterized and highly specific in their adsorptive properties. Developments have occurred in the following areas.

Purity. Because of the undesirable influence of impurities on adsorption behavior, greater attention is being given to chemical purity. This attention has taken the forms of improved process control in the production of aluminum hydroxide by the Bayer process and the use of hydroxides of higher purity available from other sources such as the Ziegler process. Most manufacturers of adsorbent aluminas now offer varying grades of chemical purity for different adsorption requirements.

Transition Forms. Production techniques have been developed to obtain aluminas possessing well-defined and well-characterized transition forms such as γ- and η-Al_2O_3. This development has contributed to greater reproducibility and consistency of adsorption behavior.

Pore Structure. Techniques developed for control of pore structure have resulted in major improvements in commercial alumina adsorbents. Microporosity is controlled by varying the precursor hydroxide and kinetics of the dehydroxylation process, while macroporosity is influenced by the forming technology. Increased understanding of these phenomena has allowed engineering of pore size distribution and degree of porosity. Currently, more manufacturers are tailoring pore structures to meet specific process requirements than was the case in the past when only a few commodity grades were available. Alumina adsorbents available at present cover the pore size range of 30–1,000,000 Å in mono-, bi-, and trimodal distributions.

Surface Properties. Modification of alumina surface to enhance selective adsorption of particular compounds is an area of rapid development. As discussed in greater detail in Chapter 8, the activated alumina surface can contain a range of surface sites differing in their chemical structure and reactivity. Modification of the surface to contain a greater proportion of surface functionalities that enhance the desired separation or reaction, while reducing undesired sites, is fast developing into a science and is a powerful tool in the design of selective adsorption processes.

A common and established approach to surface modification of aluminas is control of the degree of dehydroxylation. When the hydroxyl ion concentration is controlled, the total surface Lewis and Brönsted acidity and basicity (Chapter 8) are altered to the desired levels. Depending on the level of dehydroxylation, relative concentrations of surface hydroxyl functionalities, anion vacancies, and unsaturated Al^{3+} cations are altered over a broad range.

Further enhancement of functionalities already existing on the surface can be brought about by the addition of chemical modifiers. Adsorption of acids or bases, which alter the concentration and reactivity of alumina surface hydroxyls, is a good example. This technique is more common in catalytic applications but is gaining importance in adsorption systems using adsorbent aluminas.

A third type of surface modification involves chemical treatment of the alumina surface to introduce new functionalities that can enhance adsorption of particular species of molecules. Examples include the use of modified alumina for the removal of mercury from natural gas and the permanganate impregnation of alumina for simultaneous adsorption and oxidation of selected air pollutants. In fact, a large number of catalytic uses of aluminas fall within this group.

A number of alumina forms have been shown to possess both cation- and anion-exchange properties. With chemical modification, these properties can be enhanced. A number of these aluminas also exhibit amphoteric behavior: they have been shown to possess positive surface functionalities below their isoelectric point and negative charges above. The surface charge is amenable to modification by various means, including pH change.

Adsorption of metals using aluminas is an application based on these principles. Recent work at Alcoa Laboratories has shown that a large number of metal ions can be adsorbed along with efficient regeneration. These include alkalis, alkaline earths, transition metals, and noble metals. Figure 7.9 displays solution adsorption isotherms for a number of metal ions at pH 5.0 and a temperature of 25 °C on a commercial activated alumina. The adsorption capacities determined are comparable or even better than those for ion-exchange resins commonly used for metal removal from dilute streams with below 500-ppm metal concentration. Selectivity was achieved by choice of material, suitable chemical treatment of the solution stream, and variation of processing parameters. Commercial installations of alumina adsorption towers for metal contaminants such as mercury, silver, lead, and cadmium are currently in operation both to recover these metals and to meet environmental requirements.

In summary, activated aluminas are a group of highly versatile adsorption materials. Because of the greater understanding of their properties and

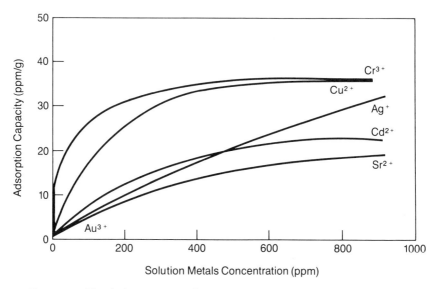

Figure 7.9. Metal absorption isotherm: nitrates (25 °C; pH 5.0; Alcoa CPN).

the ability to modify them in a desired manner, these materials have moved into applications and industries not formerly considered. In the past, petrochemical applications accounted for virtually all of the nondesiccant, adsorbent alumina market. Today this list encompasses a wide spectrum of the chemical industry including specialty chemicals, polymers, metals, waste management, and biological processing.

Analytical Methods for Activated Aluminas

A number of test methods are generally used for characterization and quality control of activated aluminas. Several other tests, specific to the end use of the product (e.g., selective adsorption and catalytic activity for particular reactions), are also applied in practice.

Moisture Content. Free moisture is the weight loss (percent) up to 300 °C. Loss on ignition (LOI) is the weight loss (percent) between 300 and 1200 °C.

Chemical Analysis. Analytical procedures applied for chemical analysis of aluminum hydroxides (Chapter 4, Analytical Procedures for Aluminum Hydroxides) are also applicable for activated aluminas. A fusion method is generally employed for bringing the alumina in solution for subsequent chemical analysis. Fusion is carried out with boric acid, a mixture of sodium carbonate and boric acid, or potassium pyrosulfate depending on the component being determined. Chemical analyses are usually carried out for

Na_2O, SiO_2, and Fe_2O_3. The soda (Na_2O) content is important for catalytic aluminas.

Phase Identification. X-ray powder diffraction methods are used to identify alumina phases, especially γ- and η-aluminas. The diffraction patterns for these two transition forms are very close, and the minor differences can only be identified by good resolution and very close examination employing comparison against well-defined standards. The use of X-ray diffraction as a process control method in the manufacture of activated aluminas has been discussed by Gerdis and Ayers (*18*).

Surface Area. Specific surface area of the product is determined by the Brunauer–Emmett–Teller nitrogen adsorption method. Several commercial instruments are available to make this determination with a high degree of automation and good reproducibility.

Pore Volume and Pore Size Distribution. Pores larger than about 30 Å are accessible by mercury penetration methods. Commercial mercury penetration porosity measurement instruments are available to determine both pore volume and pore size distribution. The volume of pores of less than 30-Å size is determined by the helium-filling method.

Mercury and Helium Density. Mercury density is determined by estimating the volume (weight) of mercury displaced by the particles and is interpreted as the average (alumina plus pores) density of the particles. This determination is carried out at about 350-mmHg vacuum, at which pressure the mercury is assumed to cover only the external surface of the particles.

A helium pycnometer is used for helium density measurements. Density is measured from helium displacement, and helium is assumed to fill all the voids present within the particles. Helium density is then considered to be the true density of the solid (alumina) mass of the particles. The total pore volume is estimated from these two density measurements by the relationship

$$\text{pore volume (cm}^3/\text{g)} = 1/\rho_{Hg} - 1/\rho_{He}$$

where ρ_{Hg} equals mercury density (g/cm^3) and ρ_{He} equals helium density (g/cm^3).

Attrition Loss. Product loss due to attrition is measured by subjecting the material to shaking under standardized conditions. A sieve shaking machine, equipped with a 200-μm opening screen, is normally used. The first 10 min of shaking removes adhering superficial dust. The weight of fines formed (expressed as a percent of the original sample) during a further fixed shaking period (10–30 min) is an indication of the attrition behavior of the product.

Crushing Strength. Two methods are used for reporting crushing strength. In the first method, usually applied to irregular granular products, a fixed weight of the product is taken in a steel cylinder equipped with a closely fitting plunger. A predetermined pressure (e.g., 50 atm) is applied to the plunger by a hydraulic cylinder, and the amount of particle breakage is determined by sieving and weighing. Results are reported as a percent of the original weight of the material.

The second procedure is applied to products of controlled geometry (e.g., spheres and tablets). In this case, the crushing strength of a single particle is determined directly from the force (or weight) required to crush the particle under compression, and this value is taken as the crushing strength of the material. The average and standard deviation of several determinations is usually reported.

Drying Capacity. The moisture removal capacity of activated alumina is of importance in gas drying applications that utilize large amounts of adsorbent aluminas.

The water removal capacity is dependent on the relative humidity of the gas to be dried, and usually moisture adsorption capacity is determined and reported over the full range of relative humidities. This measurement is carried out by equilibrating a weighed amount of the activated alumina in a constant relative humidity atmosphere. The increase in weight after reaching equilibrium conditions, shown by no further increase in weight, is measured. Constant relative humidity atmospheres can be obtained by using fixed composition solutions of H_2SO_4, H_3PO_4, or KBr in an evacuated desiccator maintained at constant temperature. A practical apparatus for carrying out this test is described by Wacks (*19*).

Literature Cited

1. Barritt, J. B. U.S. Patent 1 868 869, 1932.
2. Folkers, K.; Shovel, J. U.S. Patent 2 573 702, 1901.
3. Heftmann, E. *Chromatography*; Reinhold: New York, 1965; p 51.
4. Goodboy, K. P.; Fleming, H. L. Presented at the AIChE Annual Conference, Anaheim, CA, May 1984.
5a. Cornelius, E. B.; Milliken, T. H.; Mills, G. A.; Oblad, A. G. *J. Phys. Chem.* **1955**, *59*, 809–813.
5b. Bower, J. H. *J. Res. Nat. Bur. Stand.* **1944**, *33*, 199.
6. Stowe, J. M. *J. Phys. Chem.* **1952**, *56*, 484–489.
7. Venable, R. L.; Wade, W. H.; Hackerman, N. *J. Phys. Chem.* **1965**, *69* (1), 317–321.
8. Sanlaville, J. *Genie Chim.* **1957**, *78*, 102–122.
9. Brownell, L. *Chem. Eng. Prog.* **1950**, *46*, 415–418.

10. Derr, R. B.; Willmore, C. B. *Ind. Eng. Chem.* **1939**, *34*, 866–868.
11. Smith, W. E.; Leva, M. *Pet. Process.* **1953**, *8*, 1194–1199.
12. Rubel, F.; Woosley, R. D. "Removal of Excess Fluoride from Drinking Water"; Technical Report EPA 570/9-78-001; Office of Water Supply, EPA: Washington, DC, 1978.
13. Purushothaman, K.; Yue, C. M. "Removal of Phosphates by Sorption on Activated Alumina"; Federal Science and Technology Information PB Report 196872; U.S. Clearing House, available NTIS. 1970.
14. Gupta, S. K.; Chen, K. Y. *J.—Water Pollut. Control Fed.* **1978**, *50* (3), 493–506.
15. Alcoa; *Activated Alumina Maintenance Program—Power System Oils*; Aluminum Company of America: Pittsburgh, PA, 1955.
16. *Handbook of Chromatography*; Zweig, G.; Sherma, J., Eds. CRC Press: Cleveland, OH, 1972; Vol. 2.
17. Brockman, H.; Schodder, H. *Chem. Ber.* **1941**, *74B*, 73.
18. Gerdes, W. H.; Ayers, W. T. Presented at the Pittsburgh Conference on Analytical Chemistry and Applied Spectroscopy, Pittsburgh, PA, March 9, 1972.
19. Wacks, K. *Fette, Seifen, Anstrichm.* **1971**, *73* (1), 21–24.
20. Hiltman, L. German Olfen. DE2 434 733, 1975.
21. McKinney, C. M.; Hopkins, R. L. *Anal. Chem.* **1954**, *26* (9), 1460.
22. Tenney, H. M.; Sturgis, F. E. *Anal. Chem.* **1954**, *26* (6), 946.
23. Kumar, P.; Kunzru, D. *Ind. Chem. Man.* **1978**, No. 2.
24. Winter, G. K. German Ofen. DE2 514 018, 1975.
25. Drell, W. *Anal. Biochem.* **1970**, *34* (1), 142.
26. Graf, K.; Sauerland, H. D.; Pugen, H. German Ofen. DE2 263 992, 1974.
27. Aycock, D. F. U.S. Patent 3 943 166, 1976.
28. Porter, M. L.; Burke, J. A. *J. Assoc. Off. Anal. Chem.* **1971**, *54* (6), 1426.
29. Vishnevskii, A. V.; Martynenko, A. G.; Potashnikov, G. L. *Khim. Promst. Ser.: Promst. Gornokhim. Syr'ya* **1981**, *9*, 13.
30. Pascal, R.; Farris, C. L.; Schroepfer, G. J. *Anal. Biochem.* **1979**, *101* (1), 15.
31. Mueller, G. P. *J. Am. Ceram. Soc.* **1947**, *69*, 195.
32. Slepova, L. N.; Moldaver, B. L.; Khaletskii, A. M. *Tr. Leningr. Khim.-Farm. Inst.* **1969**, *28*, 69.
33. Laurent, C.; Billiet, H. A. H.; DeGalan, L. *Chromatographia* **1983**, *17* (5), 253.
34. Yakovleva, M. *Endokrinopatii Lech. Ikh Gorm.* **1970**, *5*, 216.
35. Bartley, I.; Hodgson, B.; Walkes, J. S.; Holme, G. *Biochem. J.* **1972**, *127* (3), 489.
36. Halpaap, H.; Reich, W.; Irmscher, K. *Testosterone, Proc. Workshop Conf., 1967*, **1968**, 25–29.
37. Luthra, M. G.; Sheltway, A. *Biochem. J.* **1972**, *126* (1), 251.

8

Catalytic Aluminas

Alumina is used in many industrial catalytic processes both as a catalyst by itself and, to a greater extent, as a support for catalytically active components. In many instances, the alumina support contributes to catalytic activity and assumes an essential role in the catalyst system. Other catalytic uses of alumina take advantage of the heat resistance and inertness of $\alpha\text{-}Al_2O_3$. Such applications include low surface area catalyst supports (e.g., steam reforming catalyst) and as an inert bed for supporting the catalyst charge in catalytic reactors.

One of the earliest examples of catalytic uses of alumina is found in alumina-catalyzed dehydration of ethyl alcohol, discovered in the years 1797–1799 (1, 2). A review of the use of alumina as a catalyst in various organic reactions was published by Sabatier (3) in 1914. Since that time, the applications of aluminas in catalytic processes have grown tremendously. The literature on this subject is extensive. A search of *Chemical Abstracts* for the years 1940–1983 turned up more than 2000 references to publications on catalytic uses of aluminas. It is beyond the purpose and scope of this book to review them all. A useful compilation, now somewhat outdated, was published by Alcoa in 1969 (4). The *Journal of Catalysis* is a rich source of scientific publications on this subject.

Although the common activated aluminas derived from Bayer hydroxide have many catalytic applications, higher purity materials (e.g., boehmite from linear alcohol production) are sometimes preferred. Because of the lower cost of the Bayer-based products, processes such as water and acid washing of the activated product have been employed to reduce the Na_2O content, which is known to have a negative influence in many catalytic processes.

Alumina-supported catalysts are prepared by impregnation and coprecipitation methods. Catalyst-forming methods include tableting, pelletizing, compacting, ball forming (in pan-type agglomerators), and extrusion. Crushing strength and resistance to attrition of the catalyst pellets are important practical considerations. Also critical are factors such as purity, surface area, pore volume, and pore size distribution that influence catalyst performance and selectivity. Although these physical aspects of alumina

0065-7719/86/0184/0133$06.00/1
© 1986 American Chemical Society

catalysts and supports are well characterized, the picture regarding surface structure and chemistry, responsible for catalytic activity, still remains unclear. This problem is the subject of considerable academic and industrial research.

Surface Structure and Catalytic Activity

Elucidation and understanding of the surface structure of alumina catalysts and supports have generally lagged behind their practical development and applications. A very limited understanding exists of the nature of the alumina surface and its interaction with the reactants and with the other catalyst components it supports.

That the surface of activated alumina shows intrinsic acidity and the catalytic activity of alumina is related to this property of the surface has long been recognized (5). Investigations of the origin and nature of the acid sites on an alumina surface are the subject of a large number of publications (6-8). Experimental results show that surface acidity increases with increasing surface dehydration and hence is influenced by the activation temperature over a broad range.

From an interpretation of infrared spectroscopic results, Peri and Hannan (7) showed that surface hydroxyl groups are formed on alumina during thermal desorption of adsorbed water. These OH^- ions on the surface behave as Brönsted acid sites. In addition, the combination of two neighboring OH^- groups to form water during the dehydration process leaves behind an exposed Al atom that, because of its electron-deficient character, behaves as a Lewis acid site. The Brönsted and Lewis acid sites have been viewed as active catalytic centers on the alumina surface. Several experimental techniques have been employed to determine the strength of these sites. The adsorption of NH_3 from the gas phase has proven to be a most promising technique; the small size of the NH_3 molecule allows it to enter very narrow pores. Infrared spectra of alumina, on which NH_3 is adsorbed, provided information about the nature of the adsorptive sites (9). Acidic character, determined from NH_3 adsorption, has been correlated with catalytic activity (8). With the removal of a greater number of OH^- ions with increasing dehydration temperature, the Brönsted acid sites, numerous at high water contents, are converted to Lewis acid sites.

That acidity could be contributed by some other structural features of the surface is demonstrated by the model proposed by Peri (9). This model, though considerably idealized, enables one to visualize the evolution (in the course of progressive dehydration) of alumina surface structure having a direct bearing on catalytic activity. The model starts with a simple, idealized representation of the γ-alumina surface shown in Figure 8.1. The proposed surface includes two outer layers of an ionic crystal of which only one face is

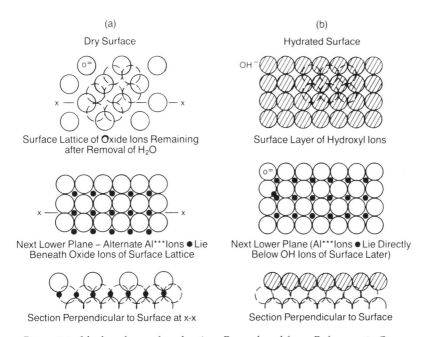

Figure 8.1. Ideal surfaces of γ-alumina. Reproduced from Reference 8. Copyright 1965 American Chemical Society.

assumed to be exposed. In the case of dry γ-Al_2O_3 the top layer contains only oxide ions (Figure 8.1a) over aluminum ions present in octahedral pattern in the next lower layer. Only half as many oxide ions are present in the top layer as in the next lower layer, which represents the (100) plane of a cubic, closely packed oxide lattice with aluminum ions located in interstices between the oxide ions. The two layers combined represent the structure of Al_2O_3. As represented in Figure 8.1b, adsorbed water is considered to convert the top layer to a filled, square lattice of hydroxyl ions. Each hydroxyl ion is assumed to be directly over an aluminum ion in the next lower layer. When the top layer is completely filled, the two upper layers correspond stoichiometrically to $Al_2O_3 \cdot H_2O$. Removal of water from the surface exposes oxide ions.

With this as a picture of the starting surface in mind, one can visualize the changes occurring on the surface with increasing degree of dehydration with increasing temperature. Computer simulation of random removal of hydroxyl pairs using a Monte Carlo technique showed that below an OH^- ion coverage of about 9.6% of the lattice, no more adjacent OH^- ion pairs are present. The state of the surface at this point is as shown in Figure 8.2. The remaining hydroxyl ions, covering about 9.6% of the surface, are found in five types of sites having from zero to four nearest oxide neighbors, as

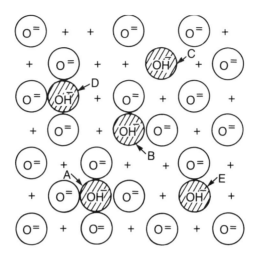

Figure 8.2. Types of isolated hydroxyl ions (+ denotes Al^{3+} in lower layer). Reproduced from Reference 8. Copyright 1965 American Chemical Society.

illustrated and identified by letter in Figure 8.2. Various types of surface defects are evident in this arrangement, where the word defect describes, in a broad sense, any irregularity in charge distribution in the surface layer. A close relationship exists between the types (Figure 8.2) of hydroxyl ions and the defects pictured by the model. Triplet defects (i.e., an oxide ion with two oxide nearest neighbors or a vacant site adjoining two vacant sites) capture protons or hydroxyl ions to form B-site hydroxyl ions. Pair defects (two oxide ions or two vacancies on immediately adjoining sites) similarly form E- or D-site hydroxyl ions.

Since no adjacent OH^- ion pairs are available at below 9.6% coverage by OH^- ions, the model, using experimental evidence, assumes migration of surface ions for further dehydration to occur. When certain rules of such migration are specified, relative changes in numerical distribution of the various hydroxyl site types could be estimated. These changes are shown in Figure 8.3. Computer simulation of the model provided reasonable interpretation and correlation of experimental observations such as infrared spectra, surface acidity, rehydration behavior, and exchange reactions of hydroxyl ions on alumina surfaces.

Of particular interest for catalytic properties is the correlation between defect structure and catalytically active sites. Of the types of defects created during dehydration, those presently thought to have the greatest catalytic importance are the triplet vacancies. These vacancy defects characteristically adjoin pair or triplet oxide defects or B-site hydroxyl ions. The triplet vacancies provide unusual exposure of the aluminum ions in the underlying layer and should constitute strong acid sites for adsorption of such electron-rich molecules as unsaturated hydrocarbons and ammonia.

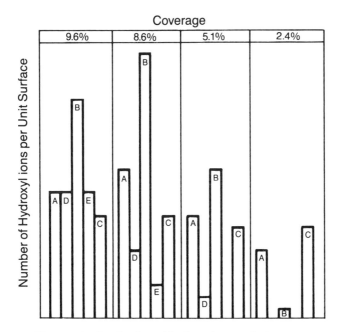

Figure 8.3. Changes in distribution of hydroxyl types during progressive removal. Reproduced from Reference 8. Copyright 1965 American Chemical Society.

Several subsequent studies and reviews have dealt with assumptions and limitations of the Peri idealized model. In an extensive study, Knözinger and Ratnasamy (11) pointed out some of the shortcomings of the Peri model and proposed a more detailed model, taking into account additional information generated during the years following Peri's publication.

The Peri model was built around the assumption that in a spinel-type alumina (γ- or η-Al_2O_3) only the (100) plane is exposed to the surface. Although this might indeed be the case for a well-defined γ-Al_2O_3, significant differences could exist in many practical alumina materials, and crystal faces other than the (100) plane might be exposed to a major extent. Knözinger's extended model is based on the assumption that either the (111) plane or, more likely, a mixture of low-index (111), (110), and (100) planes of the spinel lattice is exposed. The relative abundance of these faces is assumed to vary for different aluminas. These planes together allow five different types of OH^- groupings, the properties of which are determined by their coordination number and electrical charge. The regular lattice sites cannot be responsible for catalytic activity of alumina; catalytic activity is identified with defect sites. The existence of other types of sites is also possible depending on different energetic and geometric configurations. One exceptional defect site configuration, designated as the X site in Knözinger's model, appeared to be involved in many catalytic reactions.

The foregoing discussion is an attempt to summarize recent thinking on the nature of activated alumina surface and its relation to catalytic activity. The alumina surface is certainly complex. Although far from being complete, our understanding of its nature has been growing with the result that catalytic applications of aluminas can now be approached more scientifically and with greater confidence.

Parera (10) studied the catalytic behavior of four different commercial aluminas after activation at different temperatures. Catalytic activity for dehydration of methanol was related to acidity measurements (against an n-butylamine solution in a nonaqueous medium using a series of Hammett indicators), surface area, and phase composition (through DTA, TGA, and XRD measurements). Their conclusions generally support the surface picture derived from the Peri model. Results showed that changes in catalytic activity follow that of acidity rather than surface area. Catalytic activity and acidity both reach a maximum at the same temperature. The four products had widely different values for activity; for the same value of acidity, the catalytic activity was a characteristic of each alumina, probably related to the preparation method. However, the shapes of the curves relating activity to temperature and acidity were very similar.

The role of impurities and additives on the catalytic activity of alumina can also be interpreted by using these surface models if their effect on surface structure is considered separately from their influence on such physical aspects as surface area, pore geometry, and phase transformations.

Sodium oxide is a common impurity in alumina. As can be expected from the acid theory of catalysis by alumina, the sodium content of alumina decreases the acidity and consequently the catalytic activity in reactions where surface acid sites play a dominant role (10, 12–14). The infrared absorption study of alkali-treated aluminas by Vozdvizhenskii et al. (15) clearly shows the effect of alkali in increasing the basicity of γ-alumina. The opposite effect is observed in the presence of sulfate and other anions such as Cl, F, and SiO_2, which are generally considered to enhance the acidic nature of the alumina. Effects of Li^+, K^+, Mg^{2+}, and Fe^{3+} on properties such as surface area, pore volume, and phase composition of activated alumina catalysts have been reported by Levy and Bauer (16).

Examples of Commercial Catalytic Processes Using Alumina Catalysts

Alcohol Dehydration. Dehydration of alcohols over activated aluminas is one of the oldest catalytic processes. The products are olefins or ethers or both. Typical reaction conditions for olefin production are 300–400 °C and atmospheric pressure. Lower temperatures favor the formation of ethers. The most suitable aluminas for alcohol dehydration catalysis are those that have a large surface area (150–200 $m^2 \, g^{-1}$) with thermal and hydrothermal stability. Coke formation occurs over a period of several hundred operating hours, and the catalyst must be regenerated by burning off carbon with hot air at temperatures of 500–600 °C. The presence of alkali in the alumina has been shown to improve catalyst performance for olefin formation.

The mechanism of dehydration of alcohols over alumina catalysts has been widely studied. Pines and Pillai (17) established that the dehydration of alcohols takes place through a combined mechanism that involves an acid and a basic site. Knözinger (18) postulated that three sites comprising incompletely coordinated aluminum ions, oxygen ions, and OH groups are necessary for the dehydration of ethyl alcohol to ether. On the other hand, for methanol dehydration, the dissociative adsorption of methanol and the combination of methoxy groups for form dimethyl ether have been proposed by Parera and Figoli (19) as possible mechanisms.

Claus Catalyst. Sulfur removal is an important unit process in petroleum refining, natural gas processing, and coal gasification and liquefaction processes. Large quantities of hydrogen sulfide are produced from the desulfurization processes that are commonly used. This hydrogen sulfide is subsequently oxidized to sulfur and water in a Claus process plant. Increasingly stringent air-quality legislation is placing strong emphasis on the efficiency of Claus plants for maximum sulfur recovery and very low emission levels for sulfur compounds.

Although more expensive in first cost, activated alumina catalysts now have completely replaced bauxite in Claus converters because of their high activity, durability, and resistance to chemical, thermal, and mechanical damage.

Claus Reaction Chemistry. The operation of a typical Claus plant can be described by the basic reaction

$$2H_2S + (1 + a)O_2 \rightleftharpoons 2H_2O + [(2 - a)/n]S_n + aSO_2 \quad (1)$$

where $a = 0\text{–}2$ and $n = 2\text{–}8$.

Under typical conditions, the theoretical degree of conversion of H_2S to sulfur approaches 100% at temperatures below 250 °C. This conversion falls off rapidly, however, with increasing temperature; the conversion passes through a minimum at around 550 °C and increases again at higher temperatures.

The practical application of this reaction for low-temperature catalytic oxidation of high concentrations (greater than 25% by volume) of H_2S to sulfur has not been accomplished. The problem is due, in part, to the large exothermic heat of reaction. Reaction 1, therefore, is usually carried out in a furnace at approximately 1000 °C to produce a stoichiometric amount of SO_2 while reducing the H_2S level by nearly 75%.

$$H_2S + 3/2 O_2 \xrightarrow{\text{furnace}} SO_2 + H_2O + 124 \text{ kcal} \qquad (2)$$

The H_2S concentration is then further reduced by the reaction

$$2H_2S + SO_2 \longrightarrow 2H_2O + (3/n)S_n + 35 \text{ kcal} \qquad (3)$$

which is carried out catalytically in two or more conversion stages. Before the succeeding catalytic stages are entered and between catalytic stages, sulfur and water are condensed from the gas stream to prevent sulfur condensation on the catalyst bed and to improve the equilibrium yield. A flow sheet of the Claus plant is shown in Figure 8.4. A good description of the Claus process is given by Fischer (20).

The oxidation of H_2S by SO_2 is thermodynamically favored at low temperatures; a maximum of nearly 100% conversion to sulfur is reached at less than 150 °C. The gas-phase catalytic reaction is, however, limited by other considerations. First, the sulfur dew point is around 220 °C. Operation below this temperature results in sulfur condensation on the catalyst; the catalyst therefore becomes ineffective. A second problem is the formation of COS and CS_2. These compounds are formed in significant quantities (0.05–1.5 mol %) in the furnace by reaction of CH_4 or CO (originating from traces of organic compounds) with sulfur. These contaminants are removed in the first catalytic converter by hydrolytic decomposition.

$$COS + H_2O \longrightarrow H_2S + CO_2 \qquad (4)$$

$$CS_2 + 2H_2O \longrightarrow 2H_2S + CO_2$$

The kinetics of COS and CS_2 hydrolysis limit the operation of the first catalytic converter to temperatures between 300 and 380 °C. At this temper-

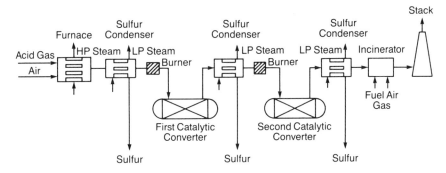

Figure 8.4. *Two-stage Claus sulfur recovery plant with auxiliary burners.*

ature (approximately 350 °C), the conversion of H_2S by reaction 3 is only 65% complete.

Subsequent lower temperature catalytic stages are used to further reduce the H_2S concentration; these stages are usually ineffective in reducing the concentration of COS and CS_2. These later catalytic stages are maintained at temperatures just above or, in many cases, below the dew point of sulfur vapor. The final stage of the Claus process is an incineration operation in which the remaining H_2S (and COS and CS_2) is oxidized with air to SO_2.

Activated Alumina Claus Catalyst. The following information describes properties and uses of activated alumina as a Claus catalyst.

PHYSICAL AND MECHANICAL PROPERTIES. These properties are listed for two commercial Claus catalysts in Table 8.I.

The spherical shape, smooth surface, and high mechanical strength of the alumina catalysts result in a lower (by nearly 30%) pressure drop than with crushed ore catalysts (e.g., bauxite) of the same particle size. Low abrasion is an important requirement in order to maintain a low pressure drop over the lifetime (up to 5 years) of the catalyst and to avoid the combination of alumina dust with molten sulfur to form a hard mass in sulfur pumps. A high crushing strength helps to minimize breakage, particularly with pneumatic or other handling during transport and converter loading.

CATALYST DEACTIVATION. Deactivation of alumina catalysts used in Claus converters is caused by several mechanisms: sulfation, thermal aging, and carbon and sulfur poisoning.

Table 8.I.
Commercial Claus Catalysts

Property	CR (Rhone Progil)	S-100 (Alcoa)
Al_2O_3 content (%)	>95.00	94.60
Total Na_2O content (%)	<0.10	0.35
Fe_2O_3 (%)	<0.05	0.05
Average water loss on ignition at 1000 °C (%)	4	5
Bulk density (g/cm³)	0.75	0.75
Surface area (m²/g)	300	300
Average total pore volume (cm³/g)	0.5	0.5
Average crushing strength for individual balls (kg)	30.0	27.2
Attrition loss (%)	<1	0.10
Shape	balls	balls
Size (mm diameter)	2–10	6

Sulfation is the formation of aluminum sulfate [$Al_2(SO_4)_3$] on the catalyst surface by reaction of alumina with SO_3 and SO_2 (which is oxidized to SO_3 and finally to sulfate during regeneration). Sulfation and resultant catalyst deactivation are favored by high levels of SO_3, the presence of free oxygen, and high SO_2 partial pressures. Sulfation is particularly harmful to the hydrolysis reactions of COS and CS_2 in the first converter.

Thermal aging of the catalyst is due to sintering, which reduces the surface area. Gradual conversion to other alumina forms is also a contributing factor. Regeneration operations accelerate the aging process.

Carbon deposition on the catalyst can usually be traced to two sources. First, high molecular weight amines such as monoethanolamine or diethanolamine are carried over from the acid gas stripping towers. This carryover causes the formation of a carbon deposit, which results in pore blockage and makes the internal catalyst surface area inaccessible. Coke formation also occurs from the cracking of heavy hydrocarbons that are present in some Claus feed streams.

Sulfur deposition on the catalyst occurs if the bed temperature falls below the sulfur dew point. The condensed sulfur fills the pores; catalytic activity is substantially decreased. This problem is often encountered when a Claus converter is shut down.

REGENERATION. Regeneration of a Claus catalyst involves removal of sulfur and burning off of carbon deposits. Sulfur is removed by passing an inert gas (e.g., N_2 or steam) while the bed temperature is maintained at 400–500 °C. All sulfur must be removed before air burning of carbon deposits commences. Sulfur has an ignition temperature of 188–243 °C, and if sulfur is not removed before air is introduced, the burning of sulfur will result in a rapid temperature rise and sulfation; catalyst activity sharply decreases. After the sulfur has been removed, air that has been diluted to 1–2% oxygen is introduced to the bed. A hot spot in excess of 200 °C is formed due to burning of carbon and moves through the bed. Dilution of the combustion air serves to limit the temperature of the hot spot by burning the carbon to CO rather than to CO_2. This reaction decreases the heat generation and controls the temperature rise.

Promoted Claus Catalysts. High surface area alumina catalysts promoted by the addition of inorganic activators have been developed for use in the first-stage Claus converters. These activators are used to improve CS_2 and COS conversions. Because most of the activators are proprietary, little published information is available on their composition. Experimental data for a lanthanum oxide promoted catalyst are presented by Baglio et al. (21). A commercially available promoted alumina catalyst is reported to contain up to 10% TiO_2. Alkali promoters (such as Na_2O) have also been used commercially.

Alumina as Catalyst Support

Alumina-supported catalysts are extensively used in the petroleum and chemical industries. In many reactions, the support itself plays a role in catalytic activity; in others, the support functions as an inert substrate for the active component. Examples of the first are mostly found in the petroleum industry. These catalysts usually have high surface areas and high porosity; the support is mostly the activated alumina type. On the other hand, many typical chemical process catalysts (e.g., ammonia synthesis, steam reforming, etc.) are characterized by lower surface areas (< 20 m^2 g^{-1}) and are nonporous or have very large diameter pores. The carrier in this case is inert. Low surface area alumina supports, with a surface area below 1 m^2 g^{-1}, consist mostly of calcined or sintered-alumina products. Intermediate surface area aluminas, in the range of 1–50 m^2 g^{-1}, contain a combination of transition alumina and α-alumina and are prepared under milder calcination conditions.

Preparation Methods. Two different methods have been commonly used for the preparation of alumina-supported catalysts: impregnation and coprecipitation.

Impregnation of the porous catalyst support with a salt solution of the active species, followed by drying and thermal decomposition of the salt, is a technique widely used for the preparation of industrial catalysts. Many of the alumina-supported catalysts are prepared in this manner. The alumina support used in the impregnation process is generally formed into its final shape (extrudates and tablets) prior to impregnation. Cervello et al. (22) reported the effect of impregnation conditions, such as solution concentration, time, previous state of support, successive impregnation, etc., on catalyst activity and selectivity.

In the coprecipitation procedure, hydroxides of aluminum together with the active component are precipitated from a salt solution of specified concentration by neutralization with ammonia or other alkali. After being washed to remove anions, the material is dried, powdered, and processed (e.g., by extrusion) to the desired shape. This procedure is followed by thermal dehydration and often a pretreatment process (e.g., reduction of metal with H_2) prior to being charged to the catalytic reactor. The coprecipitation method is usually used when more uniform distribution of the active component is desired and when the active species must be present in high concentrations.

Surface Area and Porosity. Surface area is an important property of the support because catalytic rates depend to a large extent on the amount of available active surface. However, as mentioned previously, the surface area that needs to be provided is related to the process chemistry, and a high surface area may not always be desirable. Another important property in catalytic processes is catalyst porosity and pore size distribution. Pore size is important to catalyst performance because it determines the accessibility of reactants to the active catalyst sites as well as catalyst stability, resistance to fouling, and heat transfer. Surface area and porosity can be varied independently, but only within a limited range, although possibly the same surface areas but different pore size distributions occur.

Large pores (>500 Å) in alumina supports are formed during aggregation of finer particles. Pore sizes in the range of 20–500 Å are formed in the dehydration process and are often of importance to catalyst performance. The role of micropores of <20-Å size depends on the catalytic application. In many cases micropores can provide some degree of size selectivity to molecules entering the interior of the catalyst. In other applications, particularly those involving larger molecules or processes in which carbon deposition is rapid, micropores are undesirable. Reaction products may not be able to diffuse out of the micropores and may thereby undergo undesirable secondary reactions. Hegedus (23) has given an example of modeling and optimization of pore structure and impregnation characteristics of pelleted alumina support for application in automobile emission control. The development of required pore size distribution, based on a greater understanding of reaction kinetics, has become an important goal in the preparation of catalyst supports.

Strength and Stability. Mechanical strength and long-term stability of the catalyst are primarily functions of the support. Specific requirements depend on the application. For example, for minimization of fines production, fluidized-bed operation places greater demand on the catalyst's attrition properties than does fixed-bed operation. Crushing strength is an important property because it determines the permissible depth of the catalyst bed in the reactor.

Thermal and hydrothermal stabilities are critical to many applications and determine the useful life of the catalyst. Thermal stability is an obvious requirement in high-temperature reactions. A prime example is as an automobile exhaust catalyst (*see* Automotive Exhaust Catalyst) where high resistance to thermal effects is demanded. Hydrothermal stability resistance to alteration of alumina phases under high temperature with steam present is a requirement in many catalytic processes. Catalysts that undergo frequent regeneration due to rapid coke buildup are more subject to thermal aging effects.

Examples of Catalytic Processes Using Alumina-Supported Catalysts

The technical and patent literature contains numerous examples of the use of alumina as a catalyst support. Here, the discussions will be limited to three industrially important applications that use appreciable quantities of alumina-supported catalysts. These examples also serve to illustrate the main features of the use of alumina as a catalyst support.

Chromia–Alumina Dehydrogenation Catalyst. A good example of an industrially important dehydrogenation process is the production of butadiene, used for synthetic rubber, by the catalytic dehydrogenation of *n*-butane. For this reaction, the catalyst must be sufficiently active to allow operation at low temperatures and with short contact times in order to minimize thermal cracking reactions.

The dehydrogenation of *n*-butane to butenes is endothermic. The heat of reaction is approximately 550 kcal/kg of butane. A carbon deposit forms rapidly on the catalyst during the dehydrogenation reaction. This deposit has to be removed by burning with oxygen. Consequently, the catalyst must possess a high degree of thermal stability in order to be able to resist deactivation at the temperatures attained during this burning period. Preferred catalysts for this reaction are chromia supported on alumina compositions.

Kinetics of this catalytic reaction have been studied by many workers (*24, 25*). The rate-determining step of the reaction over the chromia–alumina catalyst is suggested to be a surface reaction involving two adjacent active centers (dual site). Temperatures above 550 °C have to be maintained to obtain commercially practical conversions.

Two variations of the process, the Phillips and Houdry processes, are industrially important. Both use chromia–alumina catalysts for the dehydrogenation step but differ in operating conditions. The Houdry process catalyst is activated alumina impregnated with 18–20% chromic oxide (Cr_2O_3). Chromic acid (CrO_3) is used for impregnation of a medium pore size (50–400 Å) activated alumina with a surface area of 150–250 m^2 g^{-1}. The impregnated solid is dried at 150 °C and calcined at 550 °C in a reducing (hydrogen) atmosphere. The catalyst life is approximately 6 months. Reaction is carried out at 550 °C and a pressure below atmospheric. The reaction time is 4–10 min after which the catalyst is regenerated by burning off the carbon deposit with air preheated to 500–600 °C.

Molybdenum–Alumina Hydrorefining Catalyst. The term hydrorefining is used here in a general sense and covers hydrodesulfurization, denitrogenation, and demetallation processes. Crude desulfurization represents the largest hydrorefining operation practiced in most refineries. This process is often used as a pretreatment of feed stocks prior to other refinery operations in which the catalyst may be sulfur sensitive. The hydrorefining catalyst removes sulfur from the organic components and combines it with hydrogen to form hydrogen sulfide. The H_2S can then be separated and converted to elemental sulfur by the Claus process.

Reactions involved in denitrogenation are analogous to desulfurization conversion of organic nitrogen to ammonia, which is subsequently removed by stripping. These hydrorefining processes generally lead to demetallation and the deposition of metals as a sludge. The presence of metals usually has a deactivating influence on the catalyst used. Operations are carried out in the 300–400 °C temperature range.

The usual hydrorefining catalyst is cobalt–molybdenum on alumina. The cobalt to molybdenum ratio is normally between 1:3 to 1:4 by weight. The significant feature of these catalysts is their hydrogenation activity, which is sustained with feeds high in nitrogen and sulfur, together with their low cracking activity. In many cases, cobalt has been replaced by nickel. A high surface area (100–200 $m^2 g^{-1}$) alumina is used as a support. The metallic components of the catalyst (Mo, Co, or Ni) actually exist as the corresponding sulfides in the catalyst during use. Stabilization of the high surface area γ-alumina support with minor amounts of silica results in improved performance.

The first catalysts used in the process were produced by coprecipitation of the oxides in the presence of wet, freshly precipitated alumina gel. Later catalysts were made by impregnation of the preformed high surface area alumina carrier. Coimpregnation with solutions of cobalt nitrate and ammonium molybdate is commonly used. The metals can also be deposited by double impregnation. The support is first impregnated with the solution of one component. Then the support is dried, calcined, and impregnated with a solution of the second component, followed by normal finishing procedures. The total impregnant concentration in the final catalyst is 8–20%.

Properties of a typical desulfurization catalyst are given in Table 8.II. Apart from the high surface area, catalyst porosity is an important consideration in this application because of the nature and size of molecules that must be treated. This factor is particularly important in the case of processing of heavy residual oils that have higher metal (e.g., V and Ni) content and in which the sulfur and nitrogen compounds are more complex and difficult to process. Metal deposition on the catalyst results in loss of catalytic activ-

Table 8.II.
Properties of Typical Desulfurization Catalysts

Property	$CoMo-Al_2O_3$
Composition	
wt % of metal 1	2.5
wt % of metal 2	10.0
Physical properties	
Surface area (m^2/g)	275
Pore volume (cm^3/g)	0.6
Crushing strength (lb)	15
Abrasion loss (wt %)	2.0
Bulk density (lb/ft^3)	35
Form	extrudate
Size (in. diameter)	$1/32–1/4$

ity due to poisoning and pore plugging. A bimodal pore structure has been suggested to be a practical approach to this problem. Availability of large 600-Å macropores allows more uniform distribution of contaminants in the catalyst. These macropores are combined with small (30–100-Å) pores to give a high surface area and good activity.

A comprehensive review of hydrorefining catalysts is given by McKinley (26). Andrews (27) has presented an account of problems faced during development of a catalyst for hydrodesulfurization. Schuit and Gates (28) have given an excellent review of the chemistry and engineering of the catalytic hydrodesulfurization process.

Automotive Exhaust Catalyst. A large quantity of alumina is used as a catalyst support in automobile catalytic converters. Catalysts for these converters have evolved along two lines: monolithic and pellet forms both cylindrical and spherical. The choice between monoliths and pellets involves many technical and economic factors. The major use of alumina is as a carrier in the pellet version. The catalyst's prime function is to facilitate oxidation of hydrocarbons and CO to CO_2 and H_2O.

Typical pelleted catalysts contain platinum, palladium, and rhodium in concentrations of approximately 0.12, 0.07, and 0.009 wt %, respectively. Impregnation of the preformed support is used for depositing the metals. The structure of a General Motors pellet support is shown in Table 8.III.

Problems inherent in the development and use of catalyst supports for auto emission purification, although not very different from many industrial catalytic processes, are more difficult. Superior mechanical properties, crushing strength, and resistance to abrasion are important requirements. Furthermore, these properties should be unaffected by cyclic exposure to high temperatures (up to 1000–1100 °C) and thermal shocks. The high surface area (>60–80 m^2 g^{-1}) and high porosity (0.6–0.8 m^3 g^{-1}) must be maintained during the life of the catalyst. From weight considerations, density lowering without loss of mechanical strength is desirable. Lower density also favors faster warmup of the converter, which enables rapid attainment of performance after a cold start.

Table 8.III.
Properties of an Automobile Catalyst Support

Property	Value
Macropore volume (cm^3/g)	0.366
Micropore volume (cm^3/g)	0.627
Average large-pore radius (nm)	696
Average small-pore radius (nm)	11
Surface area (m^2/g)	114
Pellet density (g/cm^3 pellet)	0.783
Bulk density (g/cm^3 reactor)	0.48

A considerable amount of research and development work has already been done and continues to achieve these objectives (29). A paper by Gauguin et al. (30) examined the behavior of several candidate alumina carrier products with respect to thermal stability and the influence of additives on stability and durability. These researchers suggest that δ-alumina formed by the thermal dehydration sequence

$$\text{boehmite} \xrightarrow{500\ °C} \gamma\text{-Al}_2\text{O}_3 \xrightarrow{800-1000\ °C} \delta\text{-Al}_2\text{O}_3$$

possesses stable mechanical properties in the required temperature range. This transformation occurs with minimum structural modifications and hence with minimum internal strain and alteration of mechanical properties.

On the contrary, all other aluminas for which a θ phase appears in the dehydration sequence around 800 °C lose their mechanical strength. A second property affecting performance is the decrease in surface area on heating. This variation must be kept as small as possible when the product is calcined from 500 to 1000 °C. A third critical factor is the fineness and homogeneity of the precursor boehmite. Products were more attrition resistant and thermally stable as the fineness and homogeneity of the crystallites of the precursor boehmite increased.

Among additives, a positive influence was observed with some rare earth oxides (e.g., Ce) that form a more stable δ phase at higher temperatures. Both SiO_2 and Na_2O were detrimental to catalyst performance.

Literature Cited

1. Bondt, N.; Deiman, J. R.; van Troostwyr, P.; Lauwenburg, A. *Ann. Chim. Phys.* **1797**, *21*, 48.
2. Bondt, N.; Deiman, J. R.; van Troostwyr, P.; Lauwenburg, A. *Ann. Chim. Phys.* **1799**, *21*, 208.
3. Sabatier, P. *Die Katalyse in der Organischen Chemie*; Akademische Verlagsgesellschaft: Leipzig, 1914.
4. *Activated and Catalytic Aluminas*; Aluminum Company of America: Pittsburgh, PA, 1969.
5. Pines, H.; Haag, W. O. *J. Am. Chem. Soc.* **1960**, *82*, 2471.
6. Steinke, U. *Z. Anorg. Allg. Chem.* **1965**, *338*, 78.
7. Peri, J. B.; Hannan, R. B. *J. Phys. Chem.* **1960**, *64*, 1526.
8. McIver, D. S.; Tobin, H. H.; Barth, R. T. *J. Catal.* **1963**, *2*, 485.
9. Peri, J. B.; *J. Phys. Chem.* **1965**, *69*, 211–236.
10. Parera, J. M. *Ind. Eng. Chem. Prod. Res. Dev.* **1976**, *15* (4), 234–240.
11. Knözinger, H.; Ratnasamy, P. *Catal. Rev.—Sci. Eng.* **1978**, *17* (1), 31–70.
12. Parry, E. P. *J. Catal.* **1963**, *2*, 371.
13. Baily, G. C.; Holm, V. C.; Blackburn, D. M. *J. Phys. Chem.* **1958**, *62*, 1453.
14. Echigoya, W.; Shiba, T. *Bull. Tokyo Inst. Technol., Ser. B* **1960**, 133.
15. Vozdvizhenskii, V. F.; Mischenka, V. M.; Gildebrand, E. I. *Izv. Akad. Nauk Kaz. SSR, Ser. Khim.* **1977**, *27* (4), 6–11.
16. Levy, R. M.; Bauer, D. J. *J. Catal.* **1967**, *9*, 76–86.
17. Pines, H.; Pillai, C. N. *J. Am. Chem. Soc.* **1960**, *82*, 2401.

18. Knözinger, H. *Angew. Chem., Int. Ed. Engl.* **1968**, *7*, 791.
19. Parera, J. M.; Figoli, N. S. *J. Catal.* **1969**, *14*, 303.
20. Fischer, H. *Chem.-Ing.-Tech.* **1967**, *39* (9/10), 515–520.
21. Baglio, J. A.; Susa, T. J.; Wortham, D. W.; Trickett, E. A.; Lewis, T. J. *Ind. Eng. Chem. Prod. Res. Dev.* **1982**, *21*, 408–415.
22. Cervello, J.; Gracia de la Banda, J. F.; Hermana, E.; Jimenez, J. F. *Chem.-Ing.-Tech.* **1976**, *48* (6), 520–525.
23. Hegedus, L. L. *Ind. Eng. Chem. Prod. Res. Dev.* **1980**, *19*, 533–537.
24. Dodd, L. H.; Watson, K. M. *Trans. Am. Inst. Chem. Eng.* **1946**, *42*, 263.
25. Carra, S.; Forni, L.; Vintani, C. *J. Catal.* **1967**, *9*, 154–165.
26. McKinley, J. B. *Catalysis*; Emmett, P. H., Ed.; Rheinhold: New York, 1957; Vol. 5, Chapter 6.
27. Andrews, E. B. *Catalysis in Practice*; Institute of Chemical Engineers: London, 1963; pp 18–23.
28. Schuit, G. C. A.; Gates, B. C. *AIChE J.* **1973**, *19* (3), 417–420.
29. Hegedus, L. L.; Summers, J. C. U.S. Patent 4 051 073, 1977.
30. Gauguin, R.; Graulier, M.; Papée, D. In *Catalysts for the Control of Automotive Pollutants*; McEvoy, J. E., Ed.; Advances in Chemistry No. 143; American Chemical Society: Washington, DC, 1975.

9

Sodium Aluminate

Sodium aluminate is an important commercial inorganic chemical. This compound functions as an effective source of aluminum hydroxide for many industrial and technical applications. Sodium aluminate's general use as a commercial product began around 1925 when it was found to be an effective water-treatment chemical. Commercial grades of sodium aluminate are available in both solid and liquid forms. Sodium aluminate is readily soluble in water; a slight decrease in pH of dilute solutions results in voluminous precipitation of aluminum hydroxide. This compound is formed as an intermediate in the production of alumina from bauxite (Chapter 3).

Physical and Chemical Properties

Pure sodium aluminate (anhydrous) is a white crystalline solid having a formula variously given as $NaAlO_2$, $Na_2O \cdot Al_2O_3$, or $Na_2Al_2O_4$, which corresponds to equal moles of Na_2O and Al_2O_3. Commercial grades, however, always contain more than the stoichiometric amount of caustic (Na_2O); the amount of excess caustic is of the order of 1.05–1.50 times the formula requirement. Hydrated forms of sodium aluminate are crystallized from concentrated solutions. This property is shown in the Na_2O–Al_2O_3–H_2O phase diagram (Figure 9.1) at 30 °C (*1, 2*). At concentrations up to 22% by weight of Na_2O, $Al(OH)_3$ (gibbsite) is the solid phase in equilibrium with solutions of concentrations represented at the right of the solubility curve between points C and D. At concentrations between 22 and 37.5% Na_2O, monosodium aluminate hydrate ($Na_2Al_2O_4 \cdot nH_2O$) is the solid phase in equilibrium with solutions of concentrations represented at the left of the solubility curve between points D and E. At caustic concentrations above 37.5% Na_2O, a trisodium aluminate [$Na_3Al(OH)_6$] is the stable solid phase. The boundaries of the fields shown are not very precise because of solution properties that result in extremely slow attainment of equilibrium.

Detailed information regarding the different sodium aluminates crystallizing from concentrated sodium aluminate solutions is given by Ni (*3, 4*). At low temperatures (5–45 °C), monosodium aluminate hydrate of composition $Na_2O \cdot Al_2O_3 \cdot 3H_2O$ crystallizes from aluminate liquors containing

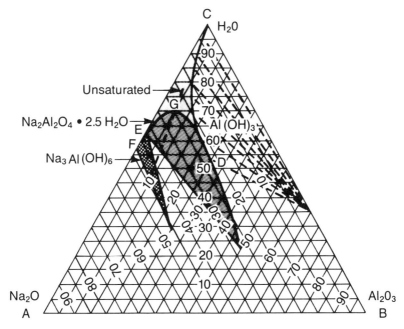

Figure 9.1. Phase diagram for the $Na_2O-Al_2O_3-H_2O$ system at 30 °C.

400–620 g of Na_2O/L. In the temperature range of 60–140 °C, an aluminate with lower water content, $2Na_2O \cdot 2Al_2O_3 \cdot 5H_2O$, crystallizes. The former compound occurs as tabular crystals with a refractive index of 1.53; the latter crystallizes as octagonal platelets with slightly higher refractive index. Anhydrous monosodium aluminate ($Na_2O \cdot Al_2O_3$) is formed at 170 °C (hydrothermal conditions) from solutions containing 530–550 g of Na_2O/L. In liquors of higher alkali concentration (750 g of Na_2O/L), this compound forms at a lower temperature (110 °C) together with trisodium aluminate. X-ray diffraction and IR spectra of the different hydrated sodium aluminates have been reported by Ni (4).

Structural and crystallographic data for solid sodium aluminate have been reported by Thery and co-workers (5, 6). Commercial grades of solid sodium aluminate contain water of hydration and excess sodium hydroxide. The X-ray diffraction pattern of the commercial products is different from that for the pure material. Typical properties of a commercial solid product are given in Table 9.I. Commercial-grade liquid sodium aluminate is a concentrated (up to 45% solids) solution of sodium aluminate in water containing excess sodium hydroxide. This compound has a $Na_2O:Al_2O_3$ molar ratio of approximately 1.15. The excess NaOH stabilizes the liquor against spontaneous precipitation of aluminum hydroxide. Some polyhydroxy organic compounds and silica have also been used to stabilize sodium aluminate solutions (7, 8).

Table 9.I.
Typical Properties and Analysis of Solid Commercial Sodium Aluminate

Property	Value
$Na_2O:Al_2O_3$ ratio	1.13–1.15
Fe (%)	0.0056
Ca and Mg (%)	0.0010
Si (%)	0.0100
Heavy metals	nil
As	nil
Hygroscopicity	slight
$NaAlO_2$ (%)	68.72
Al_2O_3 (%)	42.43
Na_2O (%)	29–30
Insolubles (%)[a]	0.1
Bulk density (lb/ft^3)	40
Absolute density (g/cm^3)	2.24
Maximum solubility (g/100 g of H_2O)[b]	60

[a] After 24 h. [b] Value was obtained at 75 °C.

Commercial Production

Small amounts of pure sodium aluminate can be conveniently prepared by fusion of equimolar quantities of sodium carbonate with aluminum acetate (5) or aluminum hydroxide.

Dissolution of aluminum hydroxides in NaOH solution is the chief commercial process for the manufacture of sodium aluminate. Aluminum hydroxide from the Bayer process can be dissolved in NaOH to obtain a solution of sodium aluminate of the required composition. Alternatively, bauxite can be used directly as the alumina source (Chapter 3). High yields can be obtained by dissolving Bayer aluminum hydroxide in 10–30% aqueous caustic at atmospheric pressure near the boiling point. Conditions used for extraction from bauxite depend on the nature of alumina minerals. Bauxites containing gibbsite are extracted at 150 °C and 4 atm of pressure, whereas boehmite-containing bauxites require higher (230 °C) temperatures and pressures. The liquid sodium aluminate obtained from the digestion process is concentrated by evaporation after separation of impurities to obtain a liquid grade. The solid product is obtained by drying the liquid. Physical characteristics of the dried product are influenced by the drying operation. Drying on a drum dryer produces a flake product. Contact with combustion products containing CO_2 during drying can result in decomposition and increased content of water-insoluble $Al(OH)_3$ in the product. A patent issued to Pevzener (9) describes a drying process that utilizes a hot gas containing no CO_2 for drying sodium aluminate in a fluidized bed. The dried product is in the form of 0.5–8-mm-size pellets, is completely soluble in water, and has a total water content of about 5 wt %.

The sinter method has also been used to produce sodium aluminate. When sodium carbonate is sintered directly with either bauxite or pure aluminum hydroxide (e.g., Bayer product) in rotary kilns at 1000 °C, an essentially anhydrous product can be obtained. When bauxite is used, the sintered mass must be water-leached and impurities must be separated to obtain a liquid sodium aluminate. Other patents and processes for producing sodium aluminate are largely variations of these basic methods. A process for the manufacture of anhydrous sodium aluminate by dissolving partially dehydrated aluminum hydroxide in molten sodium hydroxide has been patented (10).

Liquid sodium aluminate is transported in steel drums. The solid product is usually available in paper bags or fiber drums.

Analysis

Chemical analysis of commercial sodium aluminate includes the following determinations:

1. Insoluble material is determined as follows: The insoluble material remaining after extraction with aqueous caustic is filtered and washed. The residue is ignited to 1200 °C before weighing.

2. The carbonate content is determined by liberation of CO_2 on acidification and determination of CO_2 by gas buret or absorption in KOH.

3. The alumina content is determined by two methods: The gravimetric procedure involves precipitation of $Al(OH)_3$ by acid neutralization followed by ignition. Titration against HCl using the Watts and Utley (11) method is a commonly used volumetric procedure.

4. Sodium oxide is determined by the flame photometer method.

5. Soluble silica is determined by dehydration with hydrochloric or perchloric acid followed by ignition and weighing. Alternatively, the procedure followed for aluminum hydroxides (Chapter 4, Analytical Procedures for Aluminum Hydroxides) can be used for photometric determination.

Uses

The major use of sodium aluminate is in the field of water treatment, including both potable and industrial waters. One of the earliest uses of sodium aluminate was as an adjunct in the lime–soda water softening process. This application has expanded considerably over the years. Sodium aluminate increases the precipitation of hardness ions by reducing residual hardness and improves the settling and separation of suspended solids from treated water. Sodium aluminate is often used as a coagulant aid to improve flocculation efficiency of such chemicals as alum, ferric chloride, and polyelectrolytes used in the clarification of turbid waters. The formation of aluminum hydroxide flocs assists in the clarification process. Sodium aluminate has considerable affinity for dissolved silica. This property has been used to attain very low residual silica concentrations in boiler-fed water treatment. The EPA approves the use of sodium aluminate for the clarification of drinking water. The FDA approves its use in steam generation systems used for food processing. The patent literature contains examples of the use of sodium aluminate for the removal of phosphate and fluoride from industrial waste waters.

The paper industry is another large user of sodium aluminate where it is employed to improve sizing and filler retention and for pitch control in news-print mills.

Sodium aluminate is one of the principal sources of alumina for the preparation of alumina adsorbents and catalysts (Chapter 6). Its ability to react with silicates is used for the production of synthetic zeolites (molecular sieves), which have found considerable use as adsorbents and catalysts.

Many other uses of sodium aluminate have been reported. Sodium aluminate can be used as an inhibitor of glass etching by alkaline solutions; as a protector of steel surfaces during galvanizing; as an additive to improve dying, antipiling, and antistatic properties of polyester synthetic fibers; as an additive to foundry sand molds and cores to facilitate recovery of sand; as a binder in the ceramic industry; and as an additive for improving cement properties. Busler (12) gives information on the large number of patents issued on the various technical applications of sodium aluminate.

Literature Cited

1. Fricke, R.; Jucaitis, P. *Z. Anorg. Allg. Chem.* **1930**, *191*, 129–149.
2. Volf, F. F.; Kuznetsov, S. I. *J. Appl. Chem. USSR (Engl. Transl.)* **1953**, *20*, 298–302.
3. Ni, L. P. *Kompleksn. Ispol'z. Miner. Syr'ya* **1980**, *9*, 40–45.
4. Ni, L. P.; Romanov, L. G. The *Physical Chemistry of Hydrochemical Alkaline Processes for Alumina Extraction*; Nauka: Alma-Ata, USSR, 1975.
5. Thery, J.; Lejus, A. M.; Briancon, D.; Collongues, R. *Bull. Soc. Chim. Fr.* **1961**, 973.
6. Thery, J.; Briancon, D. *Rev. Hautes Temp. Refract.* **1964**, *1* (2), 221–227.
7. Lindsay, F. K.; Willey, B. F. U.S. Patent 2 345 134 to National Aluminate Corp., 1944.
8. Brown, W. H.; Armburst, B. F. Fr. Patent 1 356 638 to Reynolds Metal Co., 1964.
9. Pevzener, H. Z. U.S. Patent 4 261 958, 1978.
10. Davies, L. D. U.S. Patent 2 159 843 to Pennsylvania Salt Manufacturing Co., 1937.
11. Watts, H. L.; Utley, D. W. *Anal. Chem.* **1953**, *25*, 864–867.
12. Busler, W. R. *Kirk-Othmer-Encyclopedia of Chemical Technology*, 3rd ed.; Wiley-Interscience: New York, 1979; Vol. 2, pp 197–202.

10

Economic Data for Alumina Chemicals

Overview

Production, consumption, and cost statistics for alumina chemicals are not normally reported in trade journals and the chemical industry's technical publications. Estimates can be made by using production data for various end products that use these chemicals. Because the bulk of alumina production is used for aluminum metal smelting and alumina chemicals production is usually a side-stream operation, the major alumina producers rarely release production and cost data for chemical grades separately. The information presented in this chapter was obtained mostly from user sources, producers' commercial literature, and government agency reports and should be considered only as an overview of the economic aspects of alumina chemicals production and use.

World alumina production capacity amounted to nearly 40 million metric tons in 1982. With 80% capacity utilization, production was around 30 million tons. Alumina production has recorded an average 7–8% yearly rise in the last decade. The major part of this production (approximately 92%) was used for the smelting of aluminum metal. More than 90% of total world production came from Bayer-process refining operations; the remaining 10% is accounted for by the sinter, combination, Ziegler, and gel (neutralization of aluminum salts) processes.

The quantity of alumina used in the chemicals sector (including ceramics, refractories, and abrasives but excluding chemical uses of bauxite) in the non-Communist countries is currently (1982–1983) estimated to be 2.5 million tons per year.

Producers

Most of the industrially important alumina chemicals are produced by the major aluminum companies possessing captive alumina refining operations. The three major U.S. producers are Alcoa, Reynolds, and Kaiser. Each of these three producers operates large alumina refining plants where alumina chemicals are produced in side-stream operations. Some smaller producers have concentrated on catalytic and other specialty chemicals. Byproduct alumina (boehmite) from the Ziegler process for linear alcohol production is produced by Continental Oil Company (Conoco) in the United States.

Production of alumina chemicals by Alcoa and Reynolds is concentrated in combination-process bauxite refining plants located in Bauxite, Arkansas, and Hurricane Creek, Arkansas, respectively. Only two alumina refineries that process domestic Arkansas bauxite exist in the United States. Alcoa also has chemicals production operations at its Point Comfort (Texas) refinery and at the Vidalia (Louisiana) special products plant. Chemicals facilities of Kaiser are situated at the Baton Rouge (Louisiana) refining plant. Reynolds is reported to have ceased production operations at Hurricane Creek in 1984.

Byproduct boehmite is produced by Conoco at its linear alcohols plants located in West Germany and the United States.

European alumina chemicals production is dominated by Pechiney of France. The main production facilities are located at Salinders in France. The other important European producer is Swiss Aluminum, which owns and operates the Martinswerk GmbH alumina chemicals plant near Cologne in West Germany. Other European alumina producers, VAW and Alcoa Chemie (previously Giulini, acquired by Alcoa in 1982) in West Germany and Baco (British Aluminum) in the United Kingdom, offer many grades of alumina chemicals. The Baco plant, located at Bruntisland, is now a part of Alcan, Canada, which also has alumina chemicals production facilities in the province of Quebec. Sumitomo produces several alumina chemicals products in Japan. Showa Aluminum is another large producer of alumina chemicals in Japan.

Many other alumina plants all over the world possess the capability to dry the normal Bayer-grade hydroxide and market this product mostly for filler and aluminum chemicals production uses.

Aluminum Hydroxides

Approximately 850,000 tons of aluminum hydroxides were used in chemical applications in the United States in 1982. The major uses were as fillers (40%), for the production of aluminum chemicals (48%), and for various miscellaneous uses (12%). Bayer-process bauxite refining was the main source of this hydroxide. The whiter grades used for synthetic marble and paper industry uses came from combination-process plants. Carpet backing was the principal consumer of hydroxide filler in plastics applications followed by reinforced polyester products.

The price of aluminum hydroxides ranged from $0.20 to $0.60 per kg in 1982; the price range reflects the cost of additional processing to suit application requirements. These requirements include ground, low-iron, extra-fine, and surface-coated grades. Price of the white grades was approximately 1.5 times that of the usual off-white Bayer hydroxide. Pharmaceutical-grade gel hydroxides were at the top of the price range at nearly $0.60/kg.

Aluminum hydroxides are currently sold by all three major U.S. producers in most grades. Other firms offering aluminum hydroxide fillers probably operate only reprocessing facilities to grind or otherwise treat hydroxide obtained from the three major primary producers.

Adsorbent Activated Aluminas

U.S. production of adsorbent-grade activated aluminas, both granular and spherical, amounted to nearly 250,000 tons in 1982. Major uses were in gas drying and for humidity control.

The price of the cheaper, granular product was quoted around $0.35–0.40/kg. This Bayer crust-based product was available from two producers, Alcoa and Reynolds.

Spherical activated alumina products are obtained by the gel and fast dehydration processes. Alcoa offers products from both these processes. Kaiser's production process, originally licensed from Pechiney, follows the fast dehydration route. The price of the spherical product (in 1982) ranged from $0.55 to $0.60 per kg. Rhone-Poulonc of France is the largest seller of spherical activated alumina products in Europe. Martinswerk offers a spherical product, Granalox, produced by the fast dehydration process.

Alumina Catalysts and Catalyst Supports

The catalyst industry uses a variety of aluminas. Spherical adsorbent aluminas have been widely used as high surface area catalyst supports in organic synthesis and petroleum refining processes. Claus sulfur recovery systems are a major user of the spherical product.

Total catalytic applications of alumina in 1982 in the United States is estimated to be between 350,000 and 450,000 tons. An accurate estimate is difficult because of the variety of products, producers, and processes. Major users were Claus plants, hydrorefining processes, and automobile emission control equipment.

The price of preformed alumina for catalytic applications ranged from $0.60 to $4.00 per kg in 1982 depending on the source and purity. The higher priced products usually originated from the purer Ziegler-process boehmite. Alumina-based Claus catalyst was priced between $1.00 and $3.50 per kg.

Alumina-supported specialty catalysts are available from several manufacturers. Because of the proprietary nature of these products, little published information is available on their production volume and prices.

Sodium Aluminate

Production figures for sodium aluminate in the United States are not released. Total production in 1982 was estimated to be around 45,000 tons. The major use was in water treatment. The *Chemical Week* Directory of Chemical Suppliers lists 15 companies in the United States offering this chemical. However, primary commercial producers probably number around eight. Major producers are Nalco in the United States and Dynamit Nobel in West Germany.

Prices of bulk quantities of sodium aluminate were in the range of $0.20-0.25/kg for liquids and $0.50-0.60/kg for flake or granular solids in 1982.

INDEX

A

Activated alumina
  adsorbent applications, 107–131
  adsorbent form, 111–112
  development of adsorption
    applications, 107, 108t
  natural gas dehydration flow diagram,
    114, 115f
Activated alumina applications
  chromatography, 123–126
  gas drying, 108–119
  hydrocarbon recovery, 121
  liquid drying, 119–120
  maintenance of power system oils, 122
  water purification, 120–121
Activated alumina desiccant
  aging, 118, 119t
  characteristics, 108t, 111–112
  drying efficiency, 113
  efficiency of water removal, 109, 110f
  fouling, 118f
  gases dried, 112
  heats of wetting by water, 109–111
  industrial gas-drying operations, 113,
    114–115f
  mechanism of binding of water, 108
  natural gas drying plant, 114f
  regeneration, 116–117f, 118
  static water adsorption
    capacity, 108, 109f
Activated aluminas
  analytical methods, 128–130
  applications, 4
Activated aluminas from Bayer hydroxide
  granular activated alumina, 98
  production flow sheet, 97, 98f
  properties, 98, 99t
  uses, 98
Activated aluminas from bayerite,
  preparation, 106
Activated aluminas from boehmite,
  applications, 102
Activated aluminas from gelatinous
    aluminum hydroxide
  continuous-flow reactor, 103, 104f
  preparations, 103, 104f
  production improvements, 103, 105
  properties, 99t, 104, 105
  structure, 105f
Activated bauxites
  preparation, 97
  uses, 97

Adsorbent activated aluminas, production,
  uses, and price, 159
Adsorbent aluminas, recent developments,
  126–128
Alkalinity determination, description, 69
Alum
  background of chemical and medicinal
    usage, 1
  origin, 7
Alumen, definition, 7
Alumina
  commercial production, 31–32
  origin, 7
  production, 2
  rehydration, 82–83
  source, 2
α-Alumina, definition, 8
β-Alumina
  chemical composition, 23
  definition, 8
  preparation, 23
  properties, 23, 24t
  structure, 23, 24t
  use in the sodium–sulfur battery, 23, 25
Alumina catalysts and catalyst supports,
  production, uses, and prices, 159
Alumina chemical products, synthesis, 2, 3f
Alumina-supported catalysts
  applications, 143
  catalytic processes, 144–148
  porosity, 144
  preparation methods, 144
  stability, 144
  strength, 144
  surface area, 143–144
Alumina–water system, phase relationships,
  25, 26f, 27
Alumine, definition, 7
Aluminum fluoride, production, 65–66
Aluminum hydroxide filler
  abrasiveness, 56
  acrylic applications, 62
  advantages, 55–56
  applications, 57, 58t
  cost advantage, 56–57
  decomposition behavior, 56, 57f
  disadvantages, 56
  epoxy resin applications, 61
  fire retardancy, smoke suppression, 56
  latex applications for carpets, 58–59
  polyester system applications, 59, 60f,t,
    61f
  polyurethane applications, 62
  thermoplastic applications, 62, 63f, 64t

Aluminum hydroxides
  aluminum chemicals production, 65–67
  analytical procedures for quality control, 68–69
  bayerite, 14–16
  classification, 8, 9f
  cosmetic application, 67
  crystalline phases, 8, 9f
  filler in plastics and polymer systems, 55–63
  gelatinous hydroxides, 8, 9f
  gibbsite, 8–14
  glass industry applications, 67
  industrial applications, 55–59
  paper coating, 64–65
  paper filler, 63–64
  pharmaceutical application, 67
  production and prices, 158–159
  thermal dehydration, 74–82
  uses, 4, 158
Aluminum hydroxides and oxide
  boehmite, 17
  dehydration sequences, 76–77, 78f
  development of industrial uses, 1
  diaspore, 17
  nomenclature, 27–29
  physical properties, 8, 12t
  structures, 8, 11t
  X-ray diffraction patterns, 8, 10t
Aluminum monohydroxide, synthesis, 2, 3f
Aluminum sulfate, preparation, 65
Aluminum trihydroxide
  decomposition, 81, 82
  dehydration sequences, 74–76, 78f
  synthesis, 2, 3f
Alums, preparation, 65
Analytical methods for activated aluminas
  attrition loss, 129–130
  chemical analysis, 128–129
  crushing strength, 130
  drying capacity, 130
  mercury and helium density, 129
  moisture content, 128
  phase identification, 129
  pore volume and pore size distribution, 129
  surface area, 129
Automotive exhaust catalyst
  forms, 147
  pelleted catalysts, 147
  problems, 147–148
  properties, 147t

B

Bauxite
  components, 32
  definition, 7, 32
  mining methods, 32–33
  resources, 33
  types of ores, 32
Bayer aluminum hydroxide, source of commercial activated aluminas, 97–102
Bayer hydroxide product variants, source, 41

Bayer process
  bauxite refining plant, 36, 37f
  consumption factors for alumina production, 36, 38t
  disposal of the bauxite residue, 36
  flow sheet, 33–34, 35f, 36
  production cost factors, 36
  steps, 33, 34f, 36
Bayer-process products, properties, 36, 38, 39t
Bayerite
  dehydration sequences, 74–76, 78f
  discovery, 7
  physical properties, 12t, 16
  preparation, 14, 15f, 16
  properties, 39t, 46
  scanning electron micrograph, 46, 49f
  source of activated alumina, 105–106
  stability, 16
  structure, 11t, 16
  X-ray diffraction patterns, 10t
Boehmite
  byproduct of linear alcohol production, 48, 51t
  crystals, 17f
  dehydration mechanism, 79, 80–81f
  dehydration sequences, 76–77, 78f
  differential thermal analysis, 83, 84f
  discovery, 7
  fibrous boehmite, 48–49, 52t
  hydrothermally produced boehmite, 47–48, 50f
  physical properties, 12t, 17
  preparation, 17
  source, 17
  source of activated aluminas, 102–105
  structure, 11t, 17, 18f
  thermal dehydration effects on texture, 89
  thermogravimetric analysis, 83, 84f
Boehmite from linear alcohol production
  preparation, 48
  properties, 48, 51t

C

Calcined aluminas, uses, 4
Carbon deposition, description, 142
Catalytic aluminas
  alumina-catalyzed dehydration of ethyl alcohol, 133
  commercial catalytic processes, 138
  preparation, 133
  surface structure, 134–138
Catalytic processes using alumina-supported catalysts
  automotive exhaust catalyst, 147–148
  chromia-alumina dehydrogenation catalyst, 145
  molybdenum-alumina hydrorefining catalyst, 145–147
Catalytic processes using catalytic aluminas
  alcohol dehydration, 138–139
  Claus catalysts, 139–142
Chemical analysis, description, 69

Chromatographic separations by activated
 alumina
 adsorption activity, 124
 adsorption properties of chromatographic
  alumina, 123, 125t
 applications, 124, 126
 development in chromatographic
  techniques, 123
 examples of chemical and biological
  separations, 123, 124t
 types of aluminas, 123–124
Chromia-alumina dehydrogenation catalyst
 kinetics of catalytic reaction, 145
 Phillips and Houdry processes, 145
 production of butadiene, 145
Claus reaction chemistry
 basic reaction, 139–140
 description, 139–141
 two-stage Claus sulfur recovery plant,
  140f
Combination Bayer-sinter process
 application, 41
 steps, 40, 41f
Combustion of polymeric material, stages,
  55–56
Commercial Claus catalysts
 catalyst deactivation, 141–142
 physical and mechanical properties, 141f
 regeneration, 142
Coprecipitation procedure, description, 143
Corundum
 definition, 7
 description, 22
 physical properties, 12t, 23
 source, 22
 structure, 11t, 23
Crystalline hydroxides, classification, 8, 9f

D

Dehydration sequences
 oxide-hydroxides, 76–77, 78f
 trihydroxides, 74–76, 78f
Developments in adsorbent aluminas
 pore structure, 126
 purity, 126
 surface properties, 127, 128f
 transition forms, 126
Diaspore
 dehydration mechanism, 77, 79
 dehydration sequences, 77, 78f
 discovery, 7
 physical properties, 12t, 18
 preparation, 18
 source, 17
 structure, 11t, 18
 thermal dehydration effects on texture,
  89

E

Economic data for alumina chemicals
 absorbent activated aluminas, 159
 alumina catalysts and catalyst supports,
  159

Economic data for alumina chemicals—
 Continued
 aluminum hydroxides, 158–159
 producers, 157–158
 sodium aluminate, 160
Extra-fine hydroxide
 electron micrograph, 46, 47f
 preparation, 46
 properties, 46, 48t

F

Fibrous boehmite
 chemical compositions, 49, 52t
 physical properties, 49, 52t
 preparation, 48–49

G

Gas drying
 breakthrough capacity, 115
 by activated alumina, 108
 desiccant properties, 108–112
 industrial practice, 113, 114–115f
 split-bed design, 114–115
Gas drying by activated alumina
 desiccant aging, 118, 119t
 desiccant fouling, 118f
 regeneration, 116–117f, 118
Gelatinous aluminum hydroxide
 source of activated alumina, 103–105
 surface area vs. crystallinity, 89, 90f
Gelatinous aluminum hydroxides
 commercial production, 50–51
 development of crystal order, 20, 21f
 preparation, 19
 solubility, 19, 20f
 types of products, 19
 uses, 51–52
Gelatinous boehmite—*See* Pseudo-boehmite
Gelatinous hydroxides, classification, 8, 9f
Gibbsite
 alkali ions, 11
 crystals, 8, 13f
 dehydration sequences, 74–76, 78f
 description, 8
 differential thermal analysis, 83, 84f
 discovery, 7
 electrical energy input for dehydration,
  83–86
 optical characteristics of crystal during
  heating, 89
 physical properties, 11t
 pore structure on dehydration, 87–89
 preparation, 10–11
 production, 31
 source, 31
 structure, 11t, 13, 14f, 36, 38f
 texture development on heating, 86, 87f
 thermal decomposition, 82
 thermal dehydration, 83, 85f, 86–87, 88f
 thermogravimetric analysis, 83, 84f
 X-ray diffraction patterns, 10t
Granular activated alumina from Bayer
  plant crust
 commercial availability, 99

Granular activated alumina from Bayer
plant crust—*Continued*
  production flow sheet, 98f
  properties, 98, 99t
Granular activated alumina from
  compaction process, properties, 100
Ground hydroxides
  particle-size distributions, 43, 44t, 45f
  properties, 43, 44t

H

Hydrargillite, discovery, 7
Hydrogen recovery by activated alumina
  amount of hydrocarbons recovered, 121
  dehydration–hydrocarbon recovery unit, 121, 122f
  separation of components, 121–122
  short-cycle dehydration process, 121
Hydrothermally produced boehmite
  preparation, 47
  scanning electron micrograph, 47, 50f

I

Impregnation process, description, 143
Industrial alumina chemicals, categories, 3

L

Liquid drying by activated aluminas
  for liquid, 119–120
  regeneration, 120
Low-iron hydroxide, preparation, 43
Low-soda hydroxide, preparation, 43

M

Mercury penetration technique
  determination of pore structure, 90
  law, 90
  pore volume distributions for activated aluminas, 90, 91f
Molybdenum-alumina hydrorefining catalyst
  hydrorefining, 145–146
  preparation, 146
  properties, 146t, 147

N

Nomenclature, 27–29
Nordstrandite
  dehydration sequences, 76, 78f
  discovery, 7
  physical properties, 12t, 16
  preparation, 16
  structure, 11t, 16
Normal-grade Bayer hydroxide
  chemical composition, 39t, 42
  scanning electron micrograph, 38f, 42

Normal-grade white hydroxide, properties, 42–43

O

Oxygen index, definition, 61

P

Paper applications, aluminum hydroxide, 63–65
Particle size analysis, description, 68–69
Peri's model of surface acidity
  comparison to Knozinger's model, 137
  defect structure vs. catalytically active sites, 136
  description, 134–136
  idealized γ-alumina structure, 134, 135f
  shortcomings of model, 137
  types of isolated hydroxyl ions, 135, 136f
Phase analysis, description, 68
Physical properties, aluminum hydroxides and oxide, 8, 12t
Pore structure modification of activated aluminas
  forming procedure, 92
  incorporation of burnout materials, 92
  preparation of the hydroxide, 91–92
  sublimation, 92
Power system oil maintenance, use of activated alumina, 122–123
Production of sodium aluminate
  dissolution of aluminum hydroxides, 153
  sinter method, 154
Promoted Claus catalysts, discussion, 142
Pseudo-boehmite
  preparation, 19–21
  structure, 19–21

R

Rapid dehydration process
  description, 100
  development, 100–101
  examples of equipment, 101, 102f
  process improvements and product variations, 101
  properties of activated aluminas, 99t, 102

S

Sinter process
  application, 40
  reaction, 40
Sodium aluminate
  analysis, 154
  description, 151
  physical and chemical properties, 151–152, 153t
  preparation, 66
  production, uses, and prices, 160

Sodium aluminate—*Continued*
  sodium oxide–alumina–water phase
    diagram, 151, 152f
  transportation, 154
  uses, 154–155
Sources of transition metals
  oxidation of aluminum metal, 93
  thermal decomposition of aluminum
    salts, 93
Sulfation, description, 142
Surface modification of aluminas
  addition of chemical modifiers, 127
  applications, 127, 128f
  chemical treatment of the alumina
    surface, 127
  degree of dehydroxylation, 127
  metal adsorption, 127, 128f
Surface structure of catalytic aluminas
  effect of impurities, 138
  ideal surface of γ-alumina, 134, 135f
  nature of surface vs. catalytic activity,
    138
  Peri's model, 134–137
  surface acidity, 134
  types of isolated hydroxyl ions, 135, 136f
Surface-treated hydroxides, preparation, 46

Synthetic zeolites
  preparation, 66
  structure, 66

T

Terre argilleuse, definition, 7
Thermal aging, description, 142
Transition aluminas
  classification and nomenclature, 73–74
  formation, 75–82
  sources, 93
  X-ray powder diffraction patterns, 74, 75t

W

Water purification by activated alumina
  color and odor, 120
  fluoride, 120
  phosphate, 120–121

X

X-ray diffraction patterns, aluminum
  hydroxides and oxide, 8, 10t

Editing and indexing by Deborah H. Steiner
Production by Hilary Kanter
Jacket design by Pamela Lewis
Managing Editor: Janet S. Dodd

Typeset by Action Comp Company, Baltimore, MD, and Hot Type Ltd., Washington, DC
Printed and bound by Maple Press Company, York, PA

00 Hildebrand Hall   642.